D0074631

Clickers in Action

Increasing Student Participation in General Chemistry

Margaret R. Asirvatham
University of Colorado

Foreword
by Veronica M. Bierbaum
University of Colorado

W·W·NORTON

NEW YORK · LONDON

W. W. Norton & Company has been independent since its founding in 1923, when William Warder Norton and Mary D. Herter Norton first published lectures delivered at the People's Institute, the adult education division of New York City's Cooper Union. The Nortons soon expanded their program beyond the Institute, publishing books by celebrated academics from America and abroad. By mid-century, the two major pillars of Norton's publishing program—trade books and college texts—were firmly established. In the 1950s, the Norton family transferred control of the company to its employees and today—with a staff of four hundred and a comparable number of trade, college, and professional titles published each year—W. W. Norton & Company stands as the largest and oldest publishing house owned wholly by its employees.

© 2010 by W. W. Norton & Company, Inc.
All rights reserved.
Printed in the United States of America.

Manufacturing by Victor Graphics

Editor: Erik Fahlgren
Sciences Ancillary Editor: Matthew A. Freeman
Managing Editor, College: Marian Johnson

ISBN: 978-0-393-93353-6 (pbk.)

W. W. Norton & Company, Inc., 500 Fifth Avenue, New York, N.Y. 10110
www.wwnorton.com

W. W. Norton & Company Ltd., Castle House, 75/76 Wells Street, London W1T 3QT

1 2 3 4 5 6 7 8 9 0

Contents

FOREWORD

Although the event occurred about six years ago, I still vividly remember a defining moment in my life as a chemistry professor. I was delivering what was surely a "brilliant" lecture to my large general chemistry class, but I was deeply dismayed as I surveyed the room. There were many empty seats, and there was a constant hum of background noise as pockets of students talked and laughed among themselves. Many students were quiet, but too many of these were sleeping or enjoying the daily campus newspaper. I quickly diverted my eyes to the front rows of faithful, attentive students who are an instructor's solace and salvation. But what was wrong? Why was I failing to engage so many students with my traditional teaching approach? Fast-forward to the present semester, where I stand before a filled lecture hall of quiet, attentive students who are actively engaged in the learning process. What is the cause of this remarkable transformation?

Margaret Asirvatham has long been my close collaborator in teaching and learning, my "partner in crime" over the years, and my mentor in the instruction of large lecture classes. In Fall 2003, we co-instructed General Chemistry I to more than 900 students and, in a grand experiment, introduced Classroom Response Systems (CRS) to the course and to the Department of Chemistry and Biochemistry at the University of Colorado. Over the ensuing months and years, we have come to realize and greatly appreciate the multifaceted uses and benefits of clickers. As a superb chemical educator, Margaret has carefully documented and analyzed the uses of this new technology in the general chemistry classroom. In this book, *Clickers in Action: Increasing Student Participation in General Chemistry,* Margaret shares this wealth of information about the benefits of CRS to both students and instructors. Clickers enable students to immediately apply new concepts, to assess their understanding, and to utilize problem-solving skills. Clickers allow instructors to evaluate students' learning in real time so that misconceptions can be immediately addressed; clickers can be used to test assigned reading as an incentive for regular completion of homework; clickers can be coupled to lecture demonstrations to engage the students in both the prediction and the outcome of an experiment. Moreover, peer interaction in the use of clickers can benefit students at both extremes of the educational spectrum. The advanced student reinforces his understanding by articulating and explaining a difficult concept; the novice student often benefits from the fresh perspective provided by a peer learner. Overwhelmingly, the students have conveyed to us that they truly enjoy the challenge and the engagement of an interactive classroom. These sentiments clearly emerged from our collaborative study of student attitudes with Angela Hoekstra, a graduate student in sociology at the University of Colorado. We discovered that student acceptance of clickers was essential to the success of this intervention. Moreover, student acceptance was contingent upon their conviction that CRS is a valuable tool for teaching and learning, and not merely for attendance and grading.

Despite the myriad pedagogical benefits of CRS technology, there is an inevitable and substantial "activation energy" in implementing a new technique. How does an instructor choose and manage a system? How are questions integrated with lectures and demonstrations? How can an instructor develop a multitude of appropriate questions and answers? And perhaps most importantly, how can the requisite acceptance of clickers by students be achieved? Lack of time and resources can deter the most dedicated teacher.

In this instructor-oriented resource manual, Margaret addresses these important issues and answers these critical questions. She provides detailed guidance for the implementation and management of clickers. She discusses the wide array of clicker usage that encourages enthusiastic acceptance by the students. Moreover, she offers a treasure trove of clicker tests for twenty-three areas of general chemistry, which can be readily adapted to many levels of chemistry using a variety of textbooks.

I know that you will both enjoy and benefit from this manual. Margaret not only shares her extensive knowledge and many tricks of the trade, but she also clearly communicates her passion for teaching and her love of chemistry. I hope that this book serves as a catalyst in the transformation of your courses to include active engagement and to enhance the teaching and learning of chemistry.

Veronica M. Bierbaum
University of Colorado

PREFACE

Teaching large lectures classes has been one of the most challenging, humbling, and rewarding experiences of my academic life. We have all learned so much more, in large part due to the questions raised by our students. Unfortunately, this active discourse is limited in large classes, and I have experienced much frustration due to the impersonal nature of the teaching and learning process in traditional lecture settings. Classroom Response Systems (CRS, or "clickers") have significantly transformed teaching and learning in our large lecture courses. I am delighted to share "the agony and the ecstasy" of my experience. I designed *Clickers in Action* as an *instructor-oriented* resource (compatible with most general chemistry textbooks) to facilitate the transition from the traditional lecture to a student-centered learning environment. The use of clickers promotes interactive engagement and active participation while protecting student anonymity in the classroom. This instructional technology offers a powerful assessment tool that provides real-time feedback to both students and instructors during the course of instruction.

    Clickers in Action is *pedagogically driven* and offers instructors a large collection of class-tested resources that can be adopted and adapted to the unique needs of their students. Writing effective questions is time-intensive; this manual provides incentives to focus on conceptual understanding and reinforcement with the goal of improving knowledge retention. Questions are rated as easy, moderate, or challenging based on peer collaboration, and histograms are provided for about half the questions. The histograms were generated in large lecture courses with multiple sections often taught by the same instructor. The results are relatively similar among the sections despite the differences in composition (predominantly first-year students in the 8 A.M. section and second-year students, juniors, and seniors in the 10 and 11 A.M. sections). The histograms convey information about the concepts that were easy to comprehend as well as the difficulties encountered with some concepts and/or topics. There is plenty of room for improvement of these questions as you tailor them to the needs of your students.

    Clickers in Action is *student oriented* and offers a variety of questions to actively engage students in the learning process and to frequently solicit feedback. Some questions involve lecture demonstrations that require students to predict, observe, and explain the results. These lecture demonstrations are further enriched by the integration of questions that address macroscopic observations, symbolic representations, and atomic and molecular views of the processes involved. A section on the preparation for lecture demonstrations is included to assist instructors with time management and planning. Most students enjoy lecture demonstrations, and the effectiveness is enhanced by interactive engagement of our students using clicker questions.

    Clickers in Action addresses *atomic and molecular visualization skills* that are crucial for understanding, processing of information, and problem-solving mastery. I am happy to share pre-survey questions that we developed to gather baseline information about our students' prior knowledge and post-surveys developed to inform our teaching strategies and to assess student learning and knowledge retention. Some of the questions in this manual use atomic and molecular views with the goal of developing visualization skills and addressing common misconceptions.

    Clickers in Action brings *chemistry to life* by including questions that pertain to real-life examples such as the organic compound used as dart poison. The use of examples pertaining to the chemistry/biology interface, energy issues, and nanotechnology will get students excited about learning chemistry and motivate them to read the relevant content in their own textbooks. This is an interesting area where we can bring creativity to our classrooms as we share our passion for chemistry and science.

Clickers in Action offers a *rigorous approach*, addressing student concerns that we tend to use relatively easy quantitative problems in lecture. Some cumulative questions will combine conceptual and quantitative approaches to problem solving and a few challenging quantitative questions are included.

Clickers in Action advocates the *use of meta-communication skills* to empower students so that they appreciate the innovations in teaching. "The most powerful tool for changing students' attitudes about learning and enlisting them as active collaborators in their own education is meta-communication—high level communication about the nature and purpose of the 'normal communication' within the course" (Beatty 2004). We continually strive to encourage and enhance student buy-in, to develop pedagogically effective conceptual questions (ConcepTests[1]), and to seamlessly integrate classroom activities such as lecture demonstrations, animations, and simulations with wireless classroom response systems.

I hope that you will enjoy using this manual. Please contact me at Margaret.Asirvatham@colorado.edu to share your experiences as well as your comments, corrections, and suggestions for improvements. I am pleased to acknowledge the many dedicated individuals who have significantly contributed to my growth as a teacher and to our successful implementation of "clickers" in the large freshman general chemistry courses at CU-Boulder. The two key figures are Professors Veronica Bierbaum and Michael Dubson. Professor Bierbaum and I attended a large physics class in Spring 2003 taught by Professor Dubson, the "father of clickers," at CU-Boulder. We were stunned by the enthusiastic and vocal participation of the students as they discussed the solutions to ConcepTests, and we immediately decided to introduce clickers into our large chemistry classes. In Fall 2003, Professor Bierbaum and I team-taught the three lecture sections of General Chemistry I (she taught the first half of the semester) using infrared (IR) clickers, and we appreciated the mutual support and brainstorming sessions as we developed the first set of questions for our students. Professor Bierbaum's creativity and passion for teaching and learning science are reflected in many of the questions in this manual, and our combined enthusiasm was rewarded by acceptance of the innovation by a significant majority of our students. Professor Dubson was an excellent resource; he facilitated registration of clickers for students in several science courses and assisted us in addressing many technical and administrative issues. Two other major players, William Eberle and Robert Meyers, spent long hours perfecting the hardware and software so that Professor Bierbaum and I could enjoy teaching with clickers. The current radiofrequency (RF) technology places complete control with minimal effort in the hands of the instructor.

Several other faculty and staff members were instrumental in our successful implementation of clickers in general chemistry. Professors Barney Ellison and David Jonas contributed interesting questions to our pool of ConcepTests. We are deeply indebted to our outstanding staff members and student assistants, who often went above and beyond the call of duty, especially during the early years: Elaine Butler, Laurel Hyde, and Nathan Campbell. Dr. Laurie Langdon, Chemistry Education Specialist, arrived on the scene in Fall 2006; she played a very significant role in assisting Professor Bierbaum and me with the focus on atomic and molecular visualization. She developed some excellent chemistry concept challenge questions that required students to draw pictures or to explain visual images dealing with physical and chemical

---

[1] Eric Mazur (1997) first coined the term ConcepTest to describe conceptual clicker questions in his book *Peer Instruction: A User's Manual*, and this term is now widely recognized throughout education. Full bibliographic information can be found in Works Cited.

processes. Professor Veronica Vaida supported our initial efforts during her term as departmental chair. We are grateful to Angela Hoekstra, a doctoral student in sociology, who developed many of the student survey questions, collected responses and analyzed the data, interviewed individual students and groups, and provided valuable feedback to us. I acknowledge the support of our current faculty, Chemistry Educational Specialist, and staff, who continue to use and support clickers very effectively: Professors Robert Parson, Christine Kelly, and Susan Hendrickson; Dr. Thomas Pentecost; and Hannah Robus, Alan Foster, and Craig Cavanaugh. A very special thank-you goes to all the amazing students in General Chemistry I and II, who have taught us so much about teaching and learning.

I sincerely thank Erik Fahlgren, Editor, at W. W. Norton for encouraging me to write this manual, for his valuable suggestions and feedback, and especially for his patience. I am genuinely grateful to Matthew Freeman, Sciences Ancillaries Editor, who edited this manuscript very meticulously and provided excellent templates to maximize productivity and minimize the author's workload. And Professors Natalie Foster and Thomas Gilbert offered thoughtful comments that improved the conversational tone of this book.

Last, but not least, I am delighted to acknowledge all my family members who were very supportive and encouraging throughout this challenging experience. I especially thank my daughter, Shalini Low-Nam, and her husband, Geoffrey Graff, who combined their artistic and scientific talents to design the cover illustration.

Margaret R. Asirvatham
University of Colorado

CHAPTER 1   An Introduction to Clickers in General Chemistry

Teaching general chemistry to first-year students can be a challenging enterprise, especially in large lecture courses. It is often difficult to gauge the learning that takes place in our classrooms, despite the considerable time and effort we devote to improving our teaching. Frequently, the first meaningful feedback we receive from students is in the form of a low class average on the first hour-long exam. Education research and our own experiences confirm that the best learning does not result merely from a teacher broadcasting knowledge from the lectern, but through active student participation, with the instructor and with each other. As such, I had always wanted the ability to foster student participation and somehow receive real-time feedback about the efficacy of the activities that I developed to do so, and now with classroom response systems (CRS)—also known as "clickers"—I can. I will show you in this handbook how you too can develop and implement clickers in your own course.

What Are CRS?

What exactly is a CRS system and how does it work? A CRS system consists of two main components: a radiofrequency (RF) receiver connected to the instructor's computer that records student input, and two-way RF transmitters that students use to respond to questions or other visuals projected at the front of the classroom. The receiver unit is small enough to be portable, and typically connects to the computer via a USB cable. The transmitters are hand-held devices that look like a remote control, and, depending upon the model, come with either simple buttons labeled A–E or a fairly sophisticated alphanumeric keypad for input. Clickers get their name because students press ("click") their transmitter buttons in order to record their answers to a projected question or visual. In practical usage, the term "clicker" is often used interchangeably to refer to both the hardware system (e.g., "What clicker system do you use?") and the projected questions or visuals themselves (e.g., "I use clickers to stimulate peer discussion."). This is the convention used throughout this handbook.

With CRS systems, each student's response is recorded and is linked to a clicker ID number that is, in turn, linked with his or her university ID number. Most CRS systems will produce a histogram that displays how the class has responded, which an instructor may choose to share with the class. Any individual student's response, however, is only viewable by the instructor, not his or her peers. This is an important point: clicker response histograms allow students to see the distribution of their peers' responses and can therefore spur discussion and debate, but still grant individual anonymity. Student participation in class dramatically increases when students can respond without the fear of feeling embarrassed for answering incorrectly. I cover CRS systems in more detail in Chapter 3, "How to Choose a CRS System." The important thing to know for now is that CRS systems provide real-time feedback to the instructor about student learning and grant student anonymity, which increases class participation.

How Did I Get Started?

My introduction to CRS systems came in Spring 2003 when a colleague and I witnessed the interactive engagement and real-time feedback that clickers provide while observing Professor Michael Dubson's first-year physics class of 250 students. We were impressed by the enthusiasm of the students as they discussed and deliberated solutions to conceptual questions presented by the instructor. As we listened to the students around us, it was clear that they were actively engaged in their learning process, as new questions arose organically from these peer-led discussions. My colleague and I immediately began to investigate ways we could implement this technology in our own courses.

In Fall 2003 we introduced clickers into our large general chemistry course, which met in three sections, ranging in size from 180 to 420 students. For our first foray using clickers we decided to team-teach the course; my colleague taught all three sections for the first half of the semester, and I taught the second half. The decision to team-teach proved to be a blessing: we were partners who enthusiastically embraced the use of clicker technology and served as our own support group, determined to motivate and empower our students to be active participants in the learning process. Together we quickly learned how to engage our students in actively learning in the classroom, enable them to learn from each other, and receive important feedback in real time that could be shared with them and used to set the pace and direct progress through material in lectures.[1] We typically used three to five questions during a 50-minute lecture session. To cover the same content as we did prior to clickers, we relegated relatively easy content (as identified by clicker responses) and quantitative problem solving drill-and-practice to recitation tutorials, which were instructed by graduate teaching assistants and undergraduate learning assistants, and electronic homework. We found these practices favorable and I continue to use them in my courses.

Our Initial Results

Our first experience was a success and matched what we witnessed in Professor Dubson's lecture: that clickers help create a student-centered learning environment that increases student participation because CRS systems protect student anonymity when responding to questions. We further encouraged peer collaboration by offering bonus points for participation. Although my colleague and I were kept busy developing ConcepTests[2] throughout the semester, the brainstorming sessions and the reciprocal feedback were invigorating. While the early CRS systems my colleague and I used were not as technologically sophisticated as those currently on the market, the pedagogical benefits of using classroom response systems far outweighed the few technological problems associated with those systems.[3]

With clickers I was at last able to boost my large lecture participation *and* have quantifiable data that helped me to determine the efficacy of my teaching. No longer did I need to wonder if my time and effort in designing effective pedagogy were successful, for now I had instant feedback on each ConcepTest that either allowed me to keep the question as written or made obvious the need for refinement. This real-time feedback also gave me considerable insight into my students' understanding of previously assigned reading and of the lecture material, which enabled me to tailor content to this group of particular students. Perhaps the greatest reward of all was the buy-in by the majority of our students and their spontaneous suggestions for improvements. I will share with you the types of questions that I have found to be effective and appropriate for clicker use, based upon my six years of experience, and some data I have collected regarding student buy-in of clickers in Chapter 4, "Types of Questions and Student Surveys." I

---

[1] For more information on how clicker feedback can be used to set the pace and direction of lecture material, see Hake (1998) and Mazur (1997). Full bibliographic information can be found in Works Cited.

[2] Eric Mazur (1997) first coined the term ConcepTest and this term is now widely recognized throughout education. Full bibliographic information can be found in Works Cited.

[3] See Asirvatham (2005). Full bibliographic information can be found in Works Cited.

will teach you how to best integrate clicker questions into your lectures and share a lecture I use in Chapter 5, "Best Practices and Sample Lecture."

Challenges Unique to Teaching Chemistry
While my observation of Professor Dubson's class showed me the enthusiasm with which students in physics responded to clickers, you and I know that chemistry comes with its own unique set of challenges. The integration of macroscopic observations with atomic/molecular and symbolic representations offers challenges to experts and novices. We visualize the processes that we are discussing, and we may have even carried out experiments that present a colorful world of chemistry. Many current textbooks provide excellent atomic and molecular views of physical and chemical processes, yet students encounter difficulties with these abstract concepts and often will not look at these materials if we do not cover the information in lecture.

I was humbled by the results of the General Chemistry Concepts post-survey that was administered at my university, a week before the final exam, to 265 students after 15 weeks of instruction in first-semester general chemistry in Spring 2006. Only 13.6% of the students correctly chose (D) as the answer to the question below (Q1.1, taken from the Chemistry Concepts Inventory[4]) dealing with balanced equations and the limiting reactant concept using atomic/molecular representations. About 41% selected distractors (B) or (E) that include dimeric representations of the product! Students balance equations without too much difficulty as a mathematical exercise, yet their comprehension of the physical significance of a balanced equation is flawed. This misconception is well documented in the chemistry education research literature.[5] I assumed that my first lecture, which used atomic and molecular representations to discuss the melting of an icicle (physical change) and the electrolysis of water (chemical change), sent home a strong message about the importance of these representations in learning chemistry, but the survey gave me strong data showing that my lecture was not as effective as hoped.

Q1.1   Sample Question Testing Molecular View of a Chemical Process
This diagram represents a mixture of S atoms and O_2 molecules in a closed container.

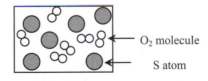

Which diagram shows the results after the mixture reacts as completely as possible according to the equation $2S + 3O_2 \rightarrow 2SO_3$?

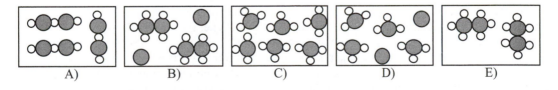

---

[4] The Chemistry Concepts Inventory is published by the *Journal of Chemical Education* and is an excellent resource. Full bibliographic information can be found in Works Cited.
[5] See Yarroch (1985) and Nurrenbern and Pickering (1987). Full bibliographic information can be found in Works Cited.

In Fall 2006, we designed and used several clicker questions that focused on atomic and molecular representations, and we used comparable questions on the hour-long exams and the final. A colleague developed Chemistry Concept Challenge Questions (CCCQs) that were assigned as weekly homework. On our first exam, 75.9% of the students (N = 792) selected the correct molecular representation of a pure compound and 90.7% (N = 679) correctly identified pure substances from atomic and molecular views on the final exam. Our results show that review and reinforcement led to some improvements in short-term and long-term understanding and knowledge retention of these chemistry concepts. The use of clickers in lectures has demonstrably helped me teach students difficult chemistry concepts, and this handbook will help you take advantage of this excellent pedagogical tool. In Chapter 2, "The Effect of Clickers on Teaching and Learning," I specifically address how clickers have been shown to improve both teacher and student performance in chemistry.

Where Do You Find the Time?
I hope the discussion thus far of clickers and how they have the ability to transform your lectures has increased your desire to use them. Obviously, lecture periods are of a set length and anything added to them means that something must be taken away. When my colleague and I first implemented clickers, we were concerned about time constraints and did our best to write questions that were designed with this awareness.

Through research and our own trial-and-error approach, we have learned that the best clicker questions are those that reinforce lecture concepts even as they test them. For example, the rules for significant figures may be covered using the ConcepTest question shown below (Q1.2). After students have responded to the question and the clicker response histogram is displayed to them, you may choose to either have a peer-led discussion of the problem or proceed and review the rules of significant figures in the context of the distractors. Thus, the clicker question does not draw away from crucial teaching time in the lecture, but rather serves as teaching time that simultaneously tests students' understanding. This is an excellent illustration of how clickers can inform you about student comprehension, provide real-time feedback, and influence the flow of your lecture.

Q1.2   Sample Question Testing Rules for Significant Figures
In which of these measured values are the zeros not significant figures?

I) 0.0591 cm   II) 504 g   III) 2.70 m   IV) 5300 L

A) I and II   B) II and III   C) I and IV   D) I, III, and IV   E) II, III, and IV

The correct answer is (C). Each measured value (I–IV) illustrates the role of the zero and why it is or is not significant. During my first experience using this clicker question, I discussed the rules prior to presenting the question, which may explain why 94% of students answered correctly.

The students had access to their lecture notes and showed confidence applying the rules correctly. Histograms, such as the one in Figure 1.1, provide useful information about student learning and attitudes. In subsequent semesters, I used this question to assess student accountability for reading the assigned material prior to class, and I reviewed the rules for significant figures after displaying the histogram.

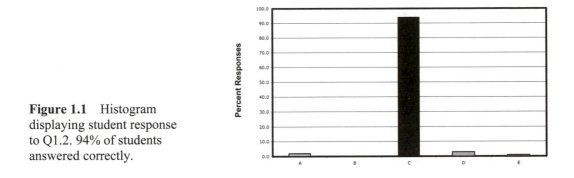

Figure 1.1 Histogram displaying student response to Q1.2. 94% of students answered correctly.

Part II of this handbook, "Questions for General Chemistry," provides ready-made clicker questions on chemistry topics covered in most first- and second-semester general chemistry courses. Each question includes comments on suggested use and my experience using it in lecture, and student response histograms gathered from my own lectures accompany about half of the questions. So that you may easily use these questions in your own lectures, all the questions in Part II are included in Microsoft PowerPoint format on the enclosed CD-ROM.

Administrative Issues

While clickers are an excellent pedagogical tool that I hope you will implement in your own lectures, in addition to learning how to use your CRS system and creating effective questions, there are some administrative issues that need to be thought through before you begin. To help make your first use of clickers as painless as possible, in Chapter 6, "Administrative Issues," I make suggestions on dealing with issues such as documenting clicker use in your course syllabus; record keeping; dealing with problems such as malfunctioning clickers, absence, and cheating; posting clicker questions online; and the pros and cons of including points for clicker use in the course grade. I hope that sharing my acquired knowledge will help you avoid some of the stumbling blocks that my colleague and I encountered when first using clickers.

Looking Ahead

This handbook is divided into three parts. In Part I, "Clickers in the Classroom," I will elaborate upon the topics introduced in this chapter, which include the proven effect of clickers on teaching and learning, how to choose a CRS system for your university, how to create the best and most effective clicker questions for integrating into your lectures, best practices for clickers, and administrative concerns. If you teach large lectures, I encourage you to read the online paper by Angel Hoekstra titled "Vibrant student voices: Exploring effects of the use of clickers in large college courses" (2008). [6] In Part II, "Questions for General Chemistry," I provide clicker questions on topics covered by most first- and second-semester general chemistry courses that are ready-made for your use. Last, in Part III, "Lecture Demonstrations," I give suggestions for how clicker use can be extended beyond assessing student comprehension of conceptual material by testing student understanding of what they witness in chemistry demonstrations of the type readily performed in lecture.

---

[6] As a doctoral student in sociology, Hoekstra worked with my colleague and me to research student attitudes toward clickers. Full bibliographic information can be found in Works Cited.

CHAPTER 2   The Effect of Clickers on Teaching and Learning

Classroom response systems (CRS) have changed the dynamics of the classroom to the benefit of both instructors and students. Instructors now receive real-time feedback about their students' conceptual understanding, especially of the macroscopic, molecular, and symbolic representations that pose a significant challenge to mastering chemistry, and students are actively engaged in the learning process through their deliberation and response to the clicker questions. CRS systems and clicker questions are ideal for classroom adoption because the current radiofrequency (RF) technology required is relatively inexpensive and user-friendly, and it provides unmatched opportunities for diagnostic and formative assessment.

In this chapter I will share with you the ways that clickers have had a positive effect on teaching and learning within my general chemistry course. I will cover how clickers have assisted me to improve formative assessment by integrating activities that influence student review and reinforcement, improve student accountability, provide data on specific individual performance relative to a class average, and make in-class surveys a viable prospect. This chapter focuses on the kind of teaching and learning results you will notice after you adopt a CRS system for use in your class. I will address how to best select a CRS system for your class in the next chapter.

How Do Clickers Affect Formative Assessment Activities?

It is widely accepted by most instructors that formative assessment occurs during a learning activity and could incorporate some diagnostic features. When used effectively, it improves learning through student-centered teaching strategies.[1] Effective formative assessment includes the identification of learning goals and the criteria for achieving them, interactive engagement between the instructor and students that provides frequent feedback so students can be accountable for their learning, and prompt attention by the instructor to difficulties encountered in the teaching and learning process. Many instructors, convinced of the benefits of formative assessment use, have already integrated formative assessment activities into their lectures, but the ability to measure, in real time, their students' learning and thereby alter the course of a lecture or further assignment has been difficult until the advent of clickers. Most CRS systems provide two primary tools that instructors can use to make a real-time assessment of their activity's effectiveness: response counters and histograms.

Many CRS systems are equipped with a counter that updates as students enter their clicker response. Watching this counter can give an instructor a good sense of how much time students need to respond. An even more powerful tool that most CRS systems provide is a histogram that displays the distribution of all given student responses. These histograms can— depending on the system—even be shared with students if an instructor chooses. Armed with the real-time information these two tools provide, an instructor is able to immediately address students' needs. If an instructor sees a histogram that either shows that a majority of students have responded incorrectly or that students took a very long time to respond to a clicker question, he or she may decide that more review or explanation are necessary and can decide to assign homework

---

[1] In their discussion of formative assessment, Black and William (1998) demonstrate that "improved formative assessment helps low achievers more than other students and so reduces the range of achievement while raising achievement overall." Full bibliographic information can be found in Works Cited.

or to review the material right then in lecture. Likewise, if a majority of students responded correctly and were very quick to respond, an instructor can tell that they have a solid grasp of the material being tested by the clicker question. In both situations, when students fail to arrive at the correct answer and when they do, clickers provide an engaged feedback cycle that benefits the instructor and student alike. Before clickers, professors would have found it very difficult to receive this type of detailed and immediate feedback assessing their students' understanding.

   Let me offer you a real-world example of how clickers affect formative assessment in general chemistry. In my own second-semester class I presented sample question Q2.1, below, at the start of lecture to test my students' comprehension of an assigned reading. I projected the question and requested that students answer on their own without consulting with their neighbors.

Q2.1   Sample Question Testing Understanding of Stepwise Ionization

Predict the relative magnitudes of the successive acid dissociation (ionization) constants for H_3PO_4 (phosphoric acid).

$$A)\ K_{a1} > K_{a2} > K_{a3} \qquad B)\ K_{a3} > K_{a2} > K_{a1}$$
$$C)\ K_{a1} > K_{a2} < K_{a3} \qquad D)\ K_{a1} < K_{a2} > K_{a3}$$

I kept the results of the response histogram to myself and did not reveal the correct answer, (A), right away. Instead, I asked students to discuss with their peers this question and its possible solution. I then had students reanswer the question after their peer deliberation. For this second round of responses I chose to display the histogram for my students to see.

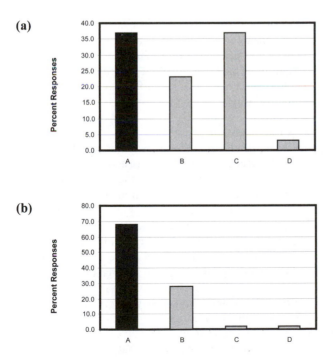

(a)

(b)

Figure 2.1 Histograms showing student response to Q2.1. Part (a) shows the distribution of student responses before discussion with their peers, and part (b) shows the distribution of student responses after peer deliberation.

As you can see from Figure 2.1, student collaboration in this instance led to a significant increase in the percentage of students who were able to identify the correct answer. Prior to peer deliberation, only 37% of students were able to select (A) as the correct answer (Figure 2.1a), whereas that percentage climbed to 68% afterward (Figure 2.1b). It is interesting to note the equal number of responses to distractors (A) and (C) when students worked on their own, and the significant decrease after peer collaboration in the number of responses to distractor (C). After displaying the histogram in Figure 2.1b, I wrote out the equations for the successive ionization steps for H_3PO_4 and then the solution (I sometimes invite students to present the solution to their peers). I have found it is very important to discuss the process and strategies involved in getting to the correct answer, even when 90% of the students are correct, in order to reinforce the conceptual understanding.

While I did not display the histogram in Figure 2.1a to the class, it demonstrated to me that many of my students either did not complete the reading assignment before coming to class or needed extra guidance. Each student knew how he or she personally responded to the question before and after deliberation. By displaying the histogram in Figure 2.1b to the class, students were able to see, on an individual basis, the benefit of reading or seeking out additional assistance from their classmates ahead of the lecture. From this example, I hope you can see how data collected by CRS systems in the classroom offer a powerful formative assessment tool, and how the histograms in Figure 2.1 support the philosophy to encourage peer collaboration in the classroom and to sow seeds for the formation of learning communities. I will briefly note that as much benefit as peer collaboration has shown to provide, I acknowledge the reality that individualized performance on exams is often the largest part of a student's course grade, and I try to address this issue by occasionally asking the students to respond individually, without peer collaboration, and after discussing my motives.

Furthermore, the grading software contained on many CRS systems can generate an extensive collection of reports about individual and class performance that can be examined and researched by instructors to answer a variety of questions pertaining to teaching and learning and the pros and cons of peer collaboration. Clicker questions may be critically evaluated in regard to distractors and more effective questions may evolve. Misconceptions and difficult and challenging concept areas may be identified and addressed in novel ways.

How Do Clickers Affect Attendance and Class Participation?

While the most significant benefit of using clickers in the classroom is the way they create a dynamic learning environment that improves the effectiveness of formative assessment activities, there are also other ancillary benefits that I appreciate just as much. One of these ancillary benefits is an increase in class attendance and participation. I saw lecture attendance in my own courses rise by as much as 30% when I made responding to clicker questions a small, but significant, part of a student's overall grade. Before implementing clickers in my courses, I would typically have a 55–60% attendance rate. This attendance rate jumped to an average of 85% once I began assigning points to students' clicker responses. You can clearly see that requiring clicker use significantly affects classroom attendance and participation. I discuss the particulars of assigning a grade to clicker use in Chapter 6.

I believe it is intuitive—but readily supported by data—that by assigning points to clicker use, student attendance will increase. But not only does clicker use increase attendance, it increases class participation too. For successful implementation of clickers, students must witness the benefits of being in class and learning to recognize what they know and what they do not know. It has been my experience that students enjoy the interactivity that CRS technology provides. Students enjoy responding to a question and then immediately viewing the results histograms that CRS systems generate. This behavior, of responding to a question and then viewing immediate results, is comfortable and an expectation that mimics their use of the everyday technology in their lives. Peer collaboration on clicker solutions is a further extension of the kinds of technology-mediated social interaction that students are familiar with and offers a comfortable learning environment for many students. It has been my experience that there is very little resistance from students to the implementation of clickers in lecture, and I'm confident that you will find this to be the case with your courses.

How Do Clickers Affect Review and Reinforcement?

Review and reinforcement are one of my favorite uses of clickers throughout the semester and in sequence courses. I have learned that students generally do not respond well to being reminded that a certain concept was covered in a previous semester. Clicker questions assist me in circumnavigating this resistance by provide a nice platform to test prior knowledge and review concepts in interesting ways without raising student ire. For example, I observed that many students struggled with the concept of constitutional isomers in the section on organic chemistry. I now review constitutional isomers when discussing intermolecular forces by having students use clickers to compare the boiling points of molecules that are constitutional isomers. Because students enjoy using clickers, they more readily accept having this kind of review.

How Do Clickers Affect Student Accountability and Individual versus Class Performance?

First-year students in general chemistry classes can often benefit from improvements in their study skills, time management, and individual accountability for learning. The use of clicker questions to test assigned reading entices students to stay abreast of lecture and to get more out of attending class. Questions that test material recently covered in lecture provide incentives for students to be attentive and to participate more actively in the learning process. Visualization skills can be enhanced by using questions related to in-lecture animations and simulations. Lecture demonstrations are usually entertaining; they can also be pedagogically effective if questions are linked to predictions and explanations of observed outcomes. Activities that focus on real-world applications and careers in chemistry can have a tremendous effect on our students. Questions that engage students both as independent thinkers and as peer collaborators can provide particularly useful feedback to students and instructors.

How Do Clickers Affect In-Class Surveys to Gather Relevant Information?

Well-designed surveys can provide a wealth of information about our students, their preparation and prior knowledge, attitudes and concerns, mastery of content, and both short-term and long-term retention of discipline-specific knowledge. Chemistry educators are continuing to develop and validate survey instruments that are usually available online. Teaching takes on an added dimension as we work toward a better understanding of the challenges encountered by beginning students. Clickers facilitate the easy aggregation of survey data through most CRS systems' response histogram feature. Unlike paper surveys, which take an inordinate amount of time to distribute, to allow for student response, to collect, and then to collate these responses by hand, surveys using CRS systems are quick and easy. Instructors need only bring to lecture the prepared survey, and then the students, already acclimated to clicker use and their timings, will respond using their RF transmitters and the CRS system software collects and immediately displays the results for you! It is my own experience that surveys given using clickers have allowed me to poll my students with greater frequency throughout the semester, which often translates into more complete and valuable feedback to help me adjust my teaching to each class individually.

How Can Clickers in This Handbook Affect My Teaching and Student Learning?

Part II of this handbook, Chapters 7–28, provides ready-made clicker questions on a variety of general chemistry topics that will help you achieve all of the teaching and performance-related improvements discussed in this chapter. My hope is that you will find these clicker questions to be a valuable resource that you can use as-is or modify for your own particular needs. I encourage you to develop your own questions that link to interesting lecture demonstrations, animations, or simulations and to then share these with your colleagues and the academic chemistry community. If your CRS system has an alphanumeric keyboard, and as you become more comfortable with implementing clickers in your own course, you may consider using the alphanumeric keypad, which can open up interesting and challenging options with new types of questions, including quantitative questions where students enter their own numeric answers.

While clickers are proving to be an extremely valuable resource for instructors, they do not offer the magic bullet to concerns about science/chemistry education. Large lecture classes can be very impersonal and students feel quite alienated from the instructor, but some of the barriers to teaching and learning may be removed when clickers are used effectively. In the following graph I share with you the results of my pursuit for any correlation between the efforts to actively engage students and their performance in my first-semester general chemistry course:

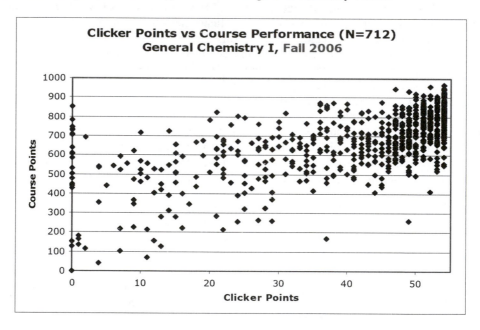

It is encouraging to observe the cluster in the top right quadrant. Perhaps students who received zero clicker points never attended class or chose not to participate in class. I believe I invested my time and effort wisely to offer my students a better learning experience! Enjoy your experience.

CHAPTER 3 How to Choose a CRS System

Perhaps you have heard horror stories about how unreliable and non-user-friendly clickers are and this has dissuaded you from implementing them in your lectures. I am here to reassure you that, while early infrared (IR) clicker systems did have their difficulties, modern radiofrequency (RF) transmitters are very easy to use, very user-friendly, and redress almost all of the shortcomings of IR systems. During my years teaching with clickers, I have used both IR and RF systems, and I can tell you with authority that the RF systems are light years ahead of their predecessors and make for an enjoyable teaching and learning experience.

In this chapter I will discuss the typical RF clicker setup, the type of hardware and software involved, the ease of use of RF systems, the associated cost of clickers, and some hints and pointers to resources that will ease and enhance your use of clickers and RF technology.

What Was So Terrible about IR Clickers and How Are RF Clickers So Much Better?
Both IR and RF systems are transmitter-receiver systems, but IR clickers, because they use infrared technology, require a clear line-of-sight between the transmitter and receiver whereas RF systems do not. Think of your television remote control. You must have a clear pathway for the signal to travel between your remote and your television in order for you to change the channel or adjust the volume. It is this requirement to have a clear line-of-sight that has been the genesis for many of the horror stories you hear about clickers. Because students sitting in a large lecture setting often did not have a direct line-of-sight between their transmitters and the receiver unit, early IR systems were prone to technical difficulties such as not registering all the students' responses. Moreover, some responses were not recorded because the IR receiver unit would become jammed with incoming signals from students trying to meet the stop time and become overloaded.

Additional problems I encountered in my early experience with IR systems included: poor transmitter design, the need for several receivers (one receiver for 250 students) that were wall-mounted and difficult to access, students being required to aim at a specific receiver, the required presence of an additional tech-savvy individual to assist with setup and operation during lecture, and exorbitant hardware costs; I spent almost $10,000 to equip three large lecture halls for clicker use! As previous chapters have illustrated, clicker technology demonstrably improves teaching and learning, and therefore I fought to overcome the hassles associated with IR technology.

Now think of a modern automobile's keyless entry system. Unlike the remote that must be pointed directly at the television to operate, you can press your car's lock/unlock clicker from almost anywhere in a parking lot and it will work. RF CRS systems, because they use a radio signal, are very much like this; they do not require line-of-sight to operate successfully and therefore have a very low failure rate. The more modern electronics inside them also ensure that they do not become overwhelmed by a large number of student responses in a very small window of time. RF systems are much more practical, more reliable, easy to use, and less expensive, making them superior to early IR systems and perfect for large lectures.

What Kind of Hardware Comprises an RF System?
The RF CRS system is comprised of two main components that must be purchased from the same supplier: a receiver that connects to a computer and the transmitters that students use to respond to

the projected clicker questions. Transmitters must be used with specific models of receivers in order to work properly. Because students often are required to purchase the transmitters as part of their class supplies, and receivers only work with their particular transmitters, it is a suggested university best practice to have all courses that use clickers use the same make and model set so that students need only purchase a transmitter once and it will work with the CRS system in all their classes.

RF receivers are mostly alike. They are small, portable, box-shaped devices that usually connect to the host computer via a USB cable. Some receivers have a USB plug built directly into them and therefore can connect directly to the host computer, thereby eliminating an additional cable, but all receivers ostensibly look and operate the same.

Transmitters, however, are where various CRS systems distinguish themselves from each other and are available in a variety of models, ranging from simple, television-remote style units with buttons labeled A–E (such as the units manufactured by i>clicker®) to those with a fairly sophisticated alphanumeric keypad, such as the Interwrite PRS® by eInstruction, shown in Figure 3.1. You may wish to start with a very simple transmitter that works well with a variety of multiple-choice and true/false questions and then, after you have become comfortable with integrating clickers in your classroom, upgrade to a system that allows students to enter numerical solutions to your questions.

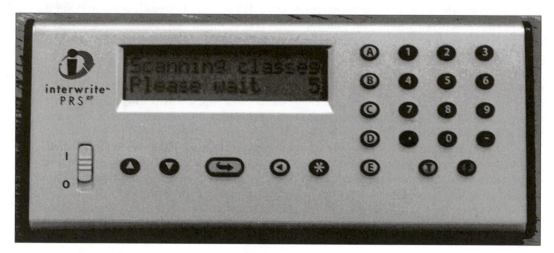

Figure 3.1   An Interwrite PRS transmitter with alphanumeric keypad.

In addition to the receiver and transmitter, you will need a video projector, all cables required to hook the receiver and projector to the same computer, and all necessary power cords. The cables needed to connect the RF receiver to the host computer are most often supplied by the CRS system vendor, along with the necessary power cords, and your university educational technology department can most likely supply you with a projector (if your lecture hall is not already media equipped) and the cable to connect the projector to the host computer.

What Kind of Software Do I Need to Operate an RF System?

Most manufacturers provide the software and software updates needed to run the RF hardware, and many do so for free. This software works as an operating system for the RF system, collects and tracks individual student responses, generates response data such as the student response counters and histograms, and integrates with familiar presentation software, including Microsoft Office.

RF system software makes data collection and tracking of individual students very easy. Each transmitter has a unique ID that can be associated with a student's university ID or some other unique identifier. Once the connection between transmitter ID and student ID is made, each response a student makes with the transmitter is recorded by the RF system, which can then provide an instructor with an individual student's performance.

Most RF system software integrates easily with popular presentation software such as Microsoft Office (including Word, PowerPoint, and Excel) and PDF files, as well as animations, simulations, and other videos. Clicker questions, data sets, and animations/simulations can be run in the applications you feel most comfortable authoring, and then the RF system software is run concurrently and integrates itself by overlaying RF system controls. When a clicker question is projected, the RF system software can take a screen capture of the question, generate and save a histogram of all student responses, and save details of individual student performance. Most RF system software contains a grading program that can generate comma separated value (CSV) or Excel files that can be exported for later inspection and use by instructors.

This software is generally very easy to use and provides a wealth of real-time and archived data that assists tremendously in the improvement of teaching and learning, especially with formative assessment activities and review and reinforcement. Clicker questions may be critically evaluated in regard to distractors and more effective questions may evolve. Misconceptions and difficult and challenging concept areas may be identified and addressed in novel ways.

How Easy Is It To Set Up and Use a Clicker System?

RF systems are now so small and portable that I carry mine along with my laptop in the bag I bring to each of my lectures. Unlike the IR systems of the past, which often required additional technical assistance, I can set up and run—and often troubleshoot—my RF system on my own. I will now describe to you the setup and use of my RF system, a system that is pretty typical of the models on the market and from which I believe you can draw a characteristic experience.

My lecture hall has a permanent video projector, so I set up the hardware in a matter of minutes. I link my laptop to the projection system via a VGA cable and then connect my laptop to the RF receiver (a single RF receiver works for up to 2,000 students) with a USB cable. With the hardware connected and powered on I confirm the RF frequency for my classroom through the RF system software. I then load my presentation software for lecture and the RF system software runs on top. I also use an overhead projector as my students appreciate the additional notes that I write down in response to student feedback during the lecture.

When I ask a clicker question, I press a button on the RF system software that is overlaid on my presentation software; this starts the recording of data. Students enter their response via their two-way RF transmitter. A vote status light on their transmitter flashes green, confirming the receiver's receipt and the recording of their response. RF technology facilitates the recording of 100 percent of the responses in a short time, and students do not have to worry about line-of-sight issues. The RF system software links responses to the clicker ID number, which is in turn linked to a student's university ID number. This allows me to track and view individual student performance later.

The RF system software provides a response counter that shows me how many students have answered my clicker question. I can then determine if I want to extend the time or close the question. The amount of time allowed for students to respond to the question depends on the type and complexity of the question. Although some instructors allow 30 seconds to a minute for some questions, this time may be too short for students to read the question and process information to arrive at the answer. I usually monitor the clock and responses displayed on the clicker counter, count down the last "30 seconds," "15 seconds," "5 seconds," and stop accepting responses when time runs out. On a few occasions, I have asked my students if more time was needed, and restarted the countdown accordingly. Students should not feel pressured to simply click in a response.

The clicker software controls float over my presentation slides, and I can easily start and stop the collection of student votes and display (or hide) the histogram. Once I have registered all student responses or closed the question, I can then display to the class a histogram of all student responses to the question, or keep this histogram hidden. The RF system software saves the histogram and associates it with the screen-captured question and all individual student response data. As I discussed in detail in Chapter 2, "The Effect of Clickers on Teaching and Learning," this real-time data collection enables me to make an on-the-spot decision about the flow of my lecture.

How Much Are RF Systems and What Features Should I Look For?

The main cost incurred to implement clickers for the first time is the price of the RF receiver unit because many vendors provide the RF system software for free, many university lecture halls are pre-wired for video projection, and students are often required to purchase the transmitter unit as part of their educational package for lecture. The range in prices can vary widely, from a few hundred dollars to a few thousand dollars, and depends largely on the complexity of the system purchased. While I will not endorse a particular model, some popular RF systems on the market today are produced by eInstruction (www.einstruction.com), Hyper-Interactive Teaching Technology (www.h-itt.com), i>clicker (www.iclicker.com), and Turning Point Technologies (www.turningtechnologies.com). Their websites are a good place to get exact details on their hardware model's capabilities and pricing. When reviewing these and other manufacturers' websites, do not let price be the sole determining factor for selecting your RF system. Other factors to consider are:

- Does the transmitter give some kind of signal to the student that the response has been received and recorded?
- How easy to install and use is the software?
- What kind of technical support does the company offer?
- What kind of presentation software does their software integrate with?
- Does the system software generate histograms? Do screen captures? Track individual student response data?

I realize that not all universities and colleges provide the same facilities to their faculty, and as such, you will most likely come up with additional purchasing criteria for your own situation. A strong suggestion that bears repeating is to keep in mind that it is a good idea to pool resources with other courses whose instructors are thinking about using clickers so that students need only purchase transmitters that work with one receiver type. Students will become resentful and be unlikely to accept clickers if they are forced to purchase multiple transmitters for different systems. Pooling resources eliminates student bitterness and makes supporting clickers easier on your university's instructional technology staff (see more discussion below).

Is There A Growing Community of Clickers Users Eager to Share Their Experience?
Yes! There is a growing community of clickers users within general chemistry that is eager to share their experience with you. Conferences and symposia usually feature a panel or two that discuss the use of CRS systems within general chemistry. These are good places to meet people to share strategies and receive help and tips about how to choose a CRS system for your course. Most manufacturers provide excellent documentation and online support for their systems, too. Some universities and colleges have adopted a single response system to maximize support for instructional technology and to minimize cost to students. Some examples of informative university websites are listed:

- Ohio State University, Turning Point Technologies, telr.osu.edu/clickers/
- Purdue University, eInstruction, www.itap.purdue.edu/tlt/einstruction/
- University of Colorado at Boulder, i>clicker, www.colorado.edu/its/cuclickers/

Experienced clicker users at large research universities have shared their combined experiences in recent publications.[1]

---

[1] See Duncan (2005) and Wieman et al. (2008). Full bibliographic information can be found in Works Cited.

CHAPTER 4 Types of Questions and Student Surveys

Classroom response systems provide opportunities for rich interactions between the instructor and students. A variety of questions can be developed to address discipline-specific content and misconceptions as well as to gather information about students' opinions, attitudes, and prior knowledge and preparation in chemistry, math, and physics. In this chapter I discuss using clickers and CRS systems to review previous lecture material and assigned reading, to reinforce conceptual understanding and visualization skills, to encourage participation in lecture demonstrations, to address problem-solving skills, to illustrate real-world chemistry applications, and to administer in-lecture student surveys.

Using Clickers to Review Previous Lecture Notes and Assigned Reading
As many of us have learned from experience, students often do not review previous lecture material or complete the assigned reading before coming to class. Using clicker questions in class to review previous lecture material or to test if students have completed the assigned reading can help us encourage the development of these essential study skills. Students can often bristle at our reminders to read ahead or review, but when clicker questions are used to quiz students about this material we can subtly influence their study behavior. Performance on these review-type questions, documented by displaying response histograms, communicates to students—without explicit instruction by the instructor—the importance of developing essential study skills.

In the example below (Q4.1), second-semester general chemistry students were asked to predict the products of the reaction using Brönsted acid-base concepts and to calculate the equilibrium constant K using information about multiple equilibria that would have been learned in their first-semester general chemistry course. This question happened to come up at the end of a lecture, so I asked students to work out an answer before the next class. I started the next lecture with this problem, and because most students did not bother to review the necessary material nor work on the assigned problem, only a small number of students (41%) responded correctly (selecting [D]). I displayed the student response histogram and used this opportunity to address several concepts related to this question. As the semester continued, student performance on review-type questions improved as they were influenced by the reminder, through the use of clicker questions, about the need to review necessary material outside of class to achieve mastery.

Q4.1   Review-Use Sample Question
Predict the magnitude of the equilibrium constant for the reaction:

$$NH_3(aq) + CH_3COOH(aq) \rightleftharpoons$$

A) $K = 0$ B) $K = 1$ C) $K \ll 1$ D) $K \gg 1$

Using acid-base concepts is a challenge for many students. I presented this same question in my first-semester organic chemistry class, and the performance was not very different from my second-semester general chemistry class.

Using Clickers to Reinforce Conceptual Understanding
Questions presented in lecture usually focus on testing student comprehension of conceptual material because students need the most work in achieving mastery in this area. Conceptual questions also often make excellent formative assessment activities, which are discussed at length in Chapter 2. The example below (Q4.2) is based on Le Châtelier's principle. This is a moderately difficult question that not only requires the application of Le Châtelier's principle, but also includes distractors that address a related concept about the equilibrium constant. This question offers you the opportunity to modify your lecture in order to cover the concept that students are struggling with, as shown by the student response histogram (Figure 4.1).

Q4.2   Conceptual Understanding Sample Question
Consider the following system at equilibrium: $SO_2(g) + Cl_2(g) \rightleftharpoons SO_2Cl_2(g)$

How will this system shift when the volume is decreased at constant temperature?
A) The position of equilibrium will remain unchanged.
B) More SO_2Cl_2 will be formed until a new position of equilibrium is attained.
C) More SO_2 will be formed until a new position of equilibrium is attained.
D) K_p will increase due to the shift in the equilibrium.
E) K_p will decrease due to the shift in the equilibrium.

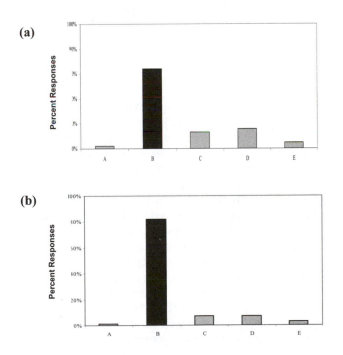

(a)

(b)

Figure 4.1   Histograms showing student response to Q4.2. Part (a) shows the student response distribution when students answered independently, and part (b) shows the student response distribution after peer discussion.

As I do with many of my formative assessment activities, I had students first respond independently and did not display the histogram. After a few minutes of deliberation with peers, I had them reanswer the question. On this question, the percent of correct responses (students selecting [B]) increased from 68% to 84%, confirming the benefits of peer collaboration (Figure 4.1).

Using Clickers to Develop and Enhance Visualization Skills

I have found that conceptual problems that test students' visualization skills are ideal questions to ask using clickers because you can render the necessary graphics or animations in a program of your choice and then insert them into your clicker system, rather than relying on crude, hand-drawn illustrations, which may easily confuse students. Question Q4.3, illustrated below, is one such question. To reinforce difficult and challenging concepts, similar questions are presented on electronic homework assignments and exams to aid in the retention of this very important chemistry skill.

Q4.3   Visualization Skill Sample Question

Which of these atomic and/or molecular views represent pure substances?

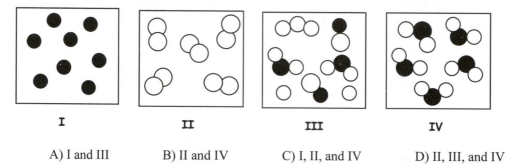

A) I and III   B) II and IV   C) I, II, and IV   D) II, III, and IV

After a class discussion of visualization of atoms, elements, and compounds, I posed this question, and 91% of the students answered correctly (selecting [C]) after peer collaboration. A similar question asked on the first hour-long exam had a 75.9% correct response and on the final exam, a 90.7% success rate. This is commendable when students are working on their own during exams, even though the concept is fairly simple, especially to experts.

Using Clickers to Encourage Participation in Lecture Demonstrations

Active participation in lecture demonstrations is a meaningful learning experience, and students often remember some experiments long after their general chemistry course. Lecture demonstrations can be very effective when students are actively engaged by appropriately designed questions that focus on observation, reflection, application of concepts learned, prediction of outcomes, and interpretation of observations and results. Using clickers is a great way to encourage student participation, and to actively involve as many students as possible.

I perform a demonstration of limiting reactants in my lecture (D4.1 below). Before I start the demonstration, I ask my students questions Q4.4–Q4.7, below, which ask them to predict limiting reactants and the outcomes of the reactions using varying stoichiometric amounts of magnesium and hydrochloric acid.

D4.1 Sample Demonstration of Limiting Reactant
(If you wish to run this demonstration in your own lecture, preparation and setup instructions are provided, beginning on page 190, in Chapter 29.)

Consider the reaction of metallic magnesium with aqueous hydrochloric acid in four different flasks.

$$Mg(s) + 2\,HCl(aq) \rightarrow MgCl_2(aq) + H_2(g)$$

Flask #	Moles of Mg(s)	Moles of HCl(aq)
1	0.0125	0.1000
2	0.0250	0.1000
3	0.0500	0.1000
4	0.1000	0.1000

Q4.4 Demonstration Question 1
Predict the limiting reactant in flask #1.
A) Mg   B) HCl   C) Both reactants are present in stoichiometric amounts

Q4.5 Demonstration Question 2
Predict the limiting reactant in flask #4.
A) Mg   B) HCl   C) Both reactants are present in stoichiometric amounts

Q4.6 Demonstration Question 3
Which flask contains stoichiometric amounts of both reactants?
A) Flask #1   B) Flask #2   C) Flask #3   D) Flask #4

Q4.7 Demonstration Question 4
What will be the relative sizes of balloons above the flasks when the reaction is complete?
A) $V_1 = V_2 = V_3 = V_4$
B) $V_1 < V_2 < V_3 < V_4$
C) $V_1 < V_2 < V_3 = V_4$
D) $V_1 < V_2 = V_3 < V_4$
E) $V_1 < V_2 = V_3 = V_4$

The selection of responses to Q4.7, a moderately difficult problem, is remarkably similar in the three different sections of first-semester chemistry that I taught in Fall 2003 (Figure 4.2). This question requires the identification of the limiting reactant and calculation of the amount of $H_2(g)$, a product of the reaction. The limiting reactant concept is one of the most challenging concepts presented in first-semester general chemistry. Peer collaboration was encouraged, and the performance was better than 80%. Students show short-term understanding and mastery (for some as a result of peer collaboration) of this concept; reinforcement of this concept throughout the semester as opportunities present themselves and to test for long-term understanding on the final exam is suggested.

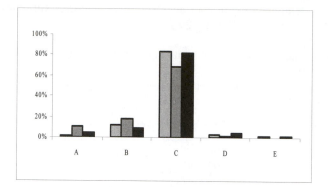

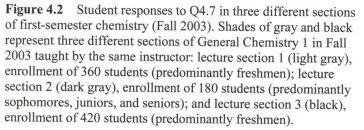

Figure 4.2   Student responses to Q4.7 in three different sections of first-semester chemistry (Fall 2003). Shades of gray and black represent three different sections of General Chemistry 1 in Fall 2003 taught by the same instructor: lecture section 1 (light gray), enrollment of 360 students (predominantly freshmen); lecture section 2 (dark gray), enrollment of 180 students (predominantly sophomores, juniors, and seniors); and lecture section 3 (black), enrollment of 420 students (predominantly freshmen).

Using Clickers to Address Problem-Solving Skills
Clicker questions do not usually address quantitative problem-solving skills, as these types of questions can be time intensive. Recitation tutorials, then, are used to focus on quantitative problem solving. Basic questions of the type shown below (Q4.8) may be presented in lecture. The use of the alphanumeric keypad (transmitter) allows students to enter their numeric answers and provides more useful feedback to the instructor.

Q4.8   Problem-Solving Skills Sample Question
If 3.423 g of aluminum sulfate is present in 1.00×10^2 mL of the solution, what is the sulfate ion concentration? Formula unit mass for $Al_2(SO_4)_3$ is 342.3 g/mol.

A) 0.100 M   B) 0.200 M   C) 0.300 M   D) 0.500 M

Calculating the concentrations of ions in solution poses a challenge to some students. It may be helpful to use cartoons to explain what happens to ionic compounds in solution and to represent the stoichiometry of the dissociation process. In my experience, a common student error is to break down polyatomic ions into smaller fragments. About 65% of students selected the correct answer, (C), to this question. A similar question using Na_3PO_4 was included on the first hour-long exam and 62.5% of students answered correctly. The ConcepTest question used in lecture was one of the questions on the final exam, and 64.9% selected the correct answer. These results confirm that students have a difficult time visualizing the ionization of salts containing polyatomic ions and/or completing the calculations.

Using Clickers to Illustrate Real-World Applications

Interesting applications of chemistry in our everyday lives can be used to captivate our students. Questions Q4.9 and Q4.10 are examples of this type of question.

Q4.9   Real-World Application Sample Question

The polymer poly(vinylidene chloride) is used for food wrap (e.g., Saran™ Wrap). Which monomer is used to make this polymer?

A) Chloroethene
B) 1,1-Dichloroethene
C) 1,2-Dichloroethene
D) 1,1,2,2-Tetrachloroethene

$$\left[\begin{array}{cccc} H & Cl & H & Cl \\ | & | & | & | \\ -C & -C & -C & -C- \\ | & | & | & | \\ H & Cl & H & Cl \end{array} \right]_n$$

The representation of the repeating unit using the expanded structural formula may explain why 89% of students responded correctly (selecting [B]) compared to only 19% in a previous semester when the condensed structural formula—$(CH_2-CCl_2-CH_2-CCl_2)_n$—was presented.

Questions that integrate applications of chemistry in life, health, and disease appeal to many biology and allied-health sciences students who comprise a significant proportion of our courses. We cover a unit on organic chemistry in our first-semester general chemistry course, following our discussion of bonding and hybridization.

Q4.10   Another Real-World Application Sample Question

Ouabain is a plant alkaloid from *Strophantus gratus* that has been used as a dart poison. It specifically binds to and inhibits sodium potassium ATPase. Carefully examine the structure shown below and determine the number of tertiary alcohol groups in the molecule.

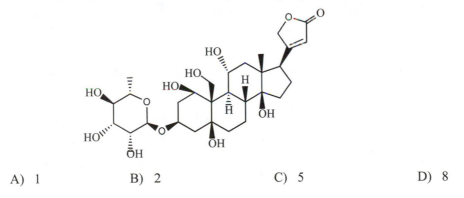

A) 1   B) 2   C) 5   D) 8

Using Clickers to Conduct In-lecture Surveys

I am always interested in gathering feedback about student opinions and attitudes throughout the course of the semester. I am also concerned about the students' preparation in math and science prior to taking my general chemistry course. CRS systems facilitate the collection of such vital information because the survey may be administered and the results collated very quickly by the CRS system software.

At my university, a doctoral sociology student was interested in investigating the relationship between learning and social interactions resulting from clicker use.[1] Because my colleague and I shared the same curiosity, we worked together and developed several questions, some of which were repeated in different forms several times during the semester, to measure this relationship. A few sample questions and the results are shown here.

This first question is related to high school chemistry preparation.

Q4.11 Survey Sample Question 1
In what year was your most recent high school chemistry course?
A) Senior
B) Junior
C) Sophomore
D) Freshman
E) Did not take HS chemistry

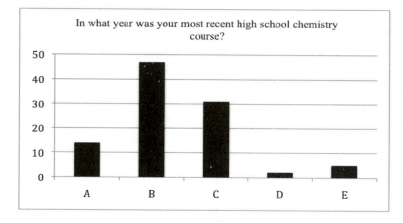

Figure 4.3 Data shown here were collected in the spring semester. Data collected in the fall often show that almost 40% take high school chemistry in their sophomore year.

Our data confirm that students generally perform better in the course if their high school chemistry course was completed in the junior or senior year of high school.

This second question illustrates student attitudes toward reading assigned sections prior to class.

Q4.12 Survey Sample Question 2
How likely are you to read the assigned reading before you come to this class?
A) Very likely
B) Somewhat likely
C) Not very likely
D) Never read before this class

---

[1] See Hoekstra (2009). Full bibliographic information can be found in Works Cited.

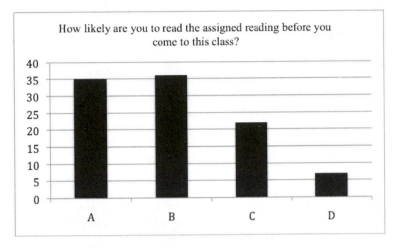

Figure 4.4 A representation of student attitudes toward reading assigned material before class. About 71% of the students polled tend to read assigned material prior to lecture.

Q4.13 Survey Sample Question 3
When clickers are used in this class, how comfortable do you feel when exchanging ideas with another student?
A) Very comfortable
B) Somewhat comfortable
C) Neither comfortable nor uncomfortable
D) Somewhat uncomfortable
E) Very uncomfortable

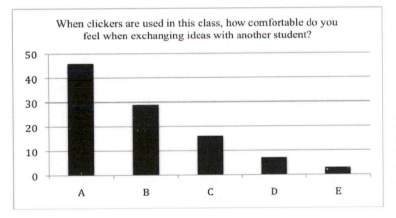

Figure 4.5 Student comfort levels about peer interaction. About 75% are comfortable about working with peers.

Although a majority of our students appreciate the use of clickers to promote interactive engagement in the classroom, a small group does not participate enthusiastically for a number of different reasons. As an instructor, I believe that the benefits of active participation far outweigh some of the perceived drawbacks, and I genuinely try to respect the views of the minority. Our goals are to continue analyzing our data to look for improvements in learning, conceptual understanding, and long-term retention, and to get more students excited about learning chemistry.

CHAPTER 5   Best Practices and Sample Lecture

It has been my experience in class that the successful implementation of interactive engagement activities is closely linked to my enthusiasm, which in turn directly affects how students embrace these new approaches to teaching and learning. Current CRS technology is so user-friendly that it is possible for instructors to feel comfortable and confident using clickers, and therefore their enthusiasm is easily transferred to their students. Additionally, effective and frequent communication is critical. We instructors need to discuss the link between instructional technology and course goals, learning goals, our teaching philosophy, and chemistry-specific issues.

To assist you in creating this link between clicker instructional technology and your course goals, below are what I believe to be "best practices" when using clickers. Much of the material here has been previously covered in this handbook, but in this chapter I present it together so that you may easily see the way that your teaching and learning strategies form a feedback loop that can give you direction for continual improvement. I have grouped the best practices into three groups: best practices to maximize clicker effectiveness for instructors, students, and lecture presentation. Furthermore, while we have previously discussed how clickers can be integrated into lecture—and Part II of this handbook contains over 250 ready-made questions for you to do this with—I present, as an example, one of my own real lectures with real clicker questions inserted in the places I use them and notes on how I utilize the real-time feedback I receive from them. If the previous chapters of this handbook can be considered clicker theory, then you should think of this chapter as clicker practice.

Best Practices

There are three components to clicker implementation: instructors, students, and the lecture presentation. I've discussed these three components in previous chapters, but here I present them together so that you may see the interplay between instructors, students, and the lecture when using clickers. I direct you to the relevant chapter that addresses a particular best practice in more detail.

Instructors: Enthusiasm and Preparation

Not only do instructors need to choose and purchase clicker systems, but to ensure successful clicker adoption they need to plan and document clicker use in their course, create and foster student buy-in, and learn to leverage the feedback they receive from their CRS systems. I encourage instructors to:

- *Be enthusiastic:* When students see that their teacher is excited to be using clickers, they will become excited too. When students view the teacher as being a lifelong learner, then they themselves will adopt this attitude. In Chapter 1, "An Introduction to Clickers in General Chemistry," I discuss how I came to use clickers and how my enthusiasm to learn and utilize clickers was adopted by my students. In my sample lecture in this chapter, I share my enthusiasm for the opportunity to combine art and science through effective visuals that teach "a picture paints a thousand words."
- *Be an effective communicator:* Effective communication aids in student buy-in. Your course syllabus must succinctly link instructional technology to your teaching

philosophy, course goals, learning goals, and chemistry-specific issues. See Chapter 6, "Administrative Issues," where I discuss documenting clicker use.

- *Do partner up:* Team-teaching a course with a partner allows you to share in "the ecstasy and the agony" of learning how to use this new technology in your classroom. Brainstorming sessions can be very productive toward this end. In Chapter 1 I discuss how having a partner helped tremendously when I first implemented clickers.

- *Motivate your students to learn:* Tell students why you use a particular lecture demonstration and then use clicker questions to actively engage them (predict outcomes and explain observations) and to collect their feedback so that you may improve the interactive activity as necessary. Use real-life examples as you design questions; many students are motivated when they make connections to other courses or a real-world application. In Chapter 4, "Types of Questions and Student Surveys," I discuss the limiting reactant concept presented with a lecture demonstration and a series of related questions. In that same chapter, I use ouabain (used as dart poison) as an interesting molecule that contains several alcohol functional groups.

- *Emphasize participation:* Foster a positive collaborative environment where students are encouraged to learn material through consulting with a peer over always feeling pressured to get a correct answer in order to boost a grade. In Chapter 2, "The Effect of Clickers on Teaching and Learning," I discuss this best practice.

- *Make clickers count:* Corollary to the above. Occasionally testing individual performance drives home that students should be actively engaged in their own learning and not rely exclusively on other's answers. This spurs students to contribute to peer discussion rather than always being passive listeners. In Chapters 2 and 6, I discuss the importance of this best practice.

- *To post clicker questions or not:* Should you post clicker questions and answers on your course website? Yes! Chapter 6 talks about how posting clicker questions and answers is a continuance of most professors' commitment to supporting student learning.

- *Solicit feedback frequently:* Gathering information on student attitudes and learning habits, especially in the initial stages of clicker implementation, reinforces to students that a willing adoption of clickers is most advantageous. Chapter 2 and Chapter 4 elaborate upon this best practice.

Students: Learning Opportunities and Peer Collaboration

Clicker use has proven to effectively increase student learning. Professor Richard E. Mayer at the University of California at Santa Barbara found that students who used clickers performed at a significantly higher level on course exams in an educational psychology class compared to an identical class using the same questions without clickers or without in-class questions.[1]

- *Address learning styles:* Clickers directly address how students learn. ConcepTests and discussions can contain visual, audio, and kinesthetic learning models that each student can identify with. In my sample lecture in this chapter, I discuss the use of column graphs and molecular/ionic scenes to address students' difficulties in understanding the similarities and differences between weak and strong acids. In addition, I write out equations and discuss with students the significance of symbolic representations. Mayer (2009) has extensively researched the "intersection of cognition, instruction, and

---

[1] See Mayer (2009). Full bibliographic information can be found in Works Cited.

technology." He advocates the use of teaching strategies that engage several different senses in the learning process.

- *Review and reinforcement:* Clicker questions allow smooth integration of review material into lecture. Students more readily accept "brushing up" on concepts when they are presented in the more palatable clicker format. Chapters 2 and 4 discuss how using clickers for review receives a positive response from students.

- *Test reading ahead:* You can subtly encourage consistent work habits by using clickers to test if students have read ahead for lecture. Chapters 2 and 4 discuss this best practice.

- *Address common student pitfalls:* Clicker flexibility allows for you to address many types of questions in lecture and to receive real-time feedback about student performance in different areas. Clickers can be used to address misconceptions, enhance conceptual understanding, develop visualization skills, and facilitate retention. Chapters 1, 2, and 4 handle addressing common student pitfalls.

- *Focus:* on conceptual understanding and problem-solving approaches. Many ConcepTests concentrate on conceptual understanding, and some types of questions are presented in Chapter 4. The instructor may model useful and effective approaches to problem solving when discussing the solution to a ConcepTest after displaying the histogram as illustrated in my sample lecture in this chapter.

- *Share strategies:* for understanding and success. Chapters 1 and 2 include examples of questions that are intended to teach topics using ConcepTests instead of presenting a list of rules. Chapter 1 has an example dealing with significant figures. Chapter 2 provides an example of engaging students, individually and as peer collaborators, in the discussion of relative ionization constants of a polyprotic acid.

- *Create incentives:* to attend and participate, conceptual questions on exams, integrate with lab wherever possible. Chapter 4 illustrates an atomic/molecular view question posed on the first hour-long exam to emphasize the importance of visualization in chemistry. Chapter 6 includes sections on attendance and participation, and Chapter 2 presents the effect of clickers on learning.

Lectures: It's Showtime!

- *Integrate:* ConcepTest questions into lecture. The sample lecture in this chapter contains a few examples of ConcepTests that are intended to assess learning in real time.

- *Make connections:* Use questions dealing with symbolic, atomic/molecular, and macroscopic connections. Chapter 1 discusses these challenges that are unique to chemistry, and question types are discussed in Chapter 4. You may wish to craft a series of questions that test cognition at these levels and guide students as they make connections and develop conceptual frameworks for future reference.

- *Discuss:* the process of thinking about the question regardless of the percentage of correct responses to the question. This information is presented in the sample lecture in this chapter.

- *Demonstrate:* Couple lecture demonstration with clicker questions to predict and explain outcomes. The sample lecture and ConcepTests are centered on a lecture demonstration; the questions focus on diagnostic and formative assessment.

- *Show:* Visuals, cartoons, animations, and simulations. Chapters 1 and 4 include sample questions with visuals. The sample lecture in this chapter includes different types of visuals that can be very effective for learning and long-term understanding.

- *Innovate:* Design innovative questions unique to the needs of your students. Adapt the 250+ class-tested questions in Part II of this handbook to your lectures and the special needs of your particular students. Easy questions may work well while the moderate and/or challenging questions may be revised as two or more questions that test fewer concepts at a time.
- *Use similar questions on quizzes and exams:* Chapter 4 includes an example of a question presented on the first hour exam.
- *Report and summarize:* Reports and summaries generated by the software may be used to address a variety of issues, especially student-centered concerns. Chapter 2 poses several questions about student learning that may be answered by examining the histograms generated by individual and/or peer performance. The reports and summaries provide a wealth of information. If accountability issues arise, you may examine the performance of an individual throughout the course. Did the student attend class? Did the student participate and benefit from peer collaboration? How did the student perform on exam questions similar to in-class ConcepTests?

Sample Lecture

We are often concerned about time constraints and losing control of the class when we attempt changes to the traditional lecture format. I have shared a sample lecture to give you an idea of how much information can be covered during a 50-minute lecture session. I enjoy the interactive format so much that I cannot imagine giving up clickers in my class, even if I had only 20–30 students!

Sample Lecture: Introduction to Acid-Base Equilibria

(ConcepTest clicker questions in this sample lecture can be found in Chapter 23 and on the enclosed CD-ROM)

Lecture Goals: To review acid-base theories and learn how to distinguish between strong and weak acids.

Comment: My students were introduced to simple acid-base concepts in high school and in first-semester general chemistry. They may even have memorized rules to recognize acids and bases and to distinguish between weak and strong acids and bases. The reading assignment for lecture required a review of these concepts. Based on my experience, I have observed a disconnect in their learning as they struggle to apply their knowledge to new scenarios. I attempt to address this disconnect using a ConcepTest for diagnostic assessment, followed by a simple lecture demonstration and a series of ConcepTests for formative assessment. How do students connect acid-base concepts and acid strength with pH?

ConcepTest

Using your prior knowledge, predict which 0.10 M aqueous solution will have the *higher* pH?
Hint: pH = $-\log [H_3O^+]$
A) HCl
B) CH_3COOH
C) Both solutions will have the same pH because the concentrations are the same.

About half the class picked distractor (C), an incorrect answer. Next, I performed the lecture demonstration and we reviewed pH concepts.

Lecture demonstration showed that the pH of a 0.10 M HCl solution is different from that of a 0.10 M CH$_3$COOH solution. Using universal indicator and the pH color chart, we found that the HCl solution had *a lower pH* and was therefore *more acidic* (since pH = −log [H$_3$O$^+$], a lower pH corresponds to a higher concentration of hydronium ions) than the CH$_3$COOH solution of the same concentration.

Class Discussion: Let us review the pH scale and extract information from this visual.

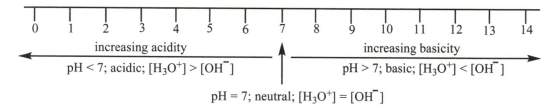

Comment: This visual evolved from real-time feedback. The goal is to connect related concepts about pH, acidity, basicity, and hydronium ion and hydroxide ion concentrations. I encourage my students to use this visual when working on problems, and have learned that some of them sketch this figure on exams to guide their thinking.

Class Discussion: Now, let us look for a rational explanation of these observations to understand why the acetic acid (CH$_3$COOH) solution has a lower [H$_3$O$^+$] than the HCl solution of the same initial concentration. We begin by reviewing theories of acids and bases.

Arrhenius Theory of Acids and Bases
Arrhenius acid: substance that produces H$^+$ ions in solution
Arrhenius bases: substance that produces OH$^-$ ions in solution

Strengths of Acids and Bases: Qualitative Predictions

Strong Acids (*You Must Know*)
Hydrohalic acids such as HI, HBr, HCl
Oxoacids such as HClO$_4$ (perchloric acid), HNO$_3$ (nitric acid), H$_2$SO$_4$ (sulfuric acid)
NOTE: The number of oxygen atoms exceeds the number of hydrogen atoms in the formula for an oxoacid by 2 or more!

Weak Acids
If you can recognize the strong acids, then you can treat all other acids (for this course) as weak acids. Some examples are HF (hydrofluoric acid), H$_3$PO$_4$ (phosphoric acid), HNO$_2$ (nitrous acid), HCN (hydrocyanic acid), and RCO$_2$H (carboxylic acids) such as acetic acid, CH$_3$CO$_2$H.

Strong Bases (*You Must Know*)
Oxides and hydroxides of alkali and alkaline earth metals

$Na_2O(s) + H_2O(l) \rightarrow 2\,NaOH(aq)$; **molecular equation**
$O^{2-}(aq) + H_2O(l) \rightarrow 2\,OH^{-}(aq)$; **net ionic equation**

Magnesium hydroxide, $Mg(OH)_2$, is a sparingly soluble salt. However, the small amount of dissolved salt behaves as a strong electrolyte. The simplified depiction shown below may aid in visualizing the solution process and the solubility equilibria of a sparingly soluble salt (solubility equilibria are discussed in detail later in the course)

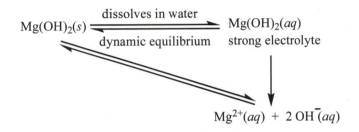

Weak Bases
Ammonia, NH_3
Amines such as CH_3NH_2 (methylamine), $(CH_3)_2NH$ (dimethylamine), and $(CH_3)_3N$ (trimethylamine) are often associated with a fishy odor. Lower molar mass amines are quite volatile. You can neutralize these weak bases with lemon juice (source of citric acid) and convert them to salts that are not volatile, thus minimizing the fishy odor! Other amines such as putrescine and cadaverine are associated with the odor of decaying flesh.

Brönsted Theory of Acids and Bases
Brönsted acid: proton (H^+) donor
Brönsted base: proton (H^+) acceptor

Comment: Use an example and identify the Brönsted acid, Brönsted base, conjugate acid and conjugate base, and discuss relationships between Brönsted acid and conjugate base (and Brönsted base and conjugate acid). The next ConcepTest is designed to help students identify HCl as a strong acid, to apply the Brönsted concept, and to visualize the species present in the 0.10 M HCl solution.

ConcepTest
What <u>major</u> species are present in an aqueous 0.10 M HCl solution?

I) HCl   II) H_2O III) H_3O^+ IV) Cl^- V) OH^-

  A) I and II   B) II, III, and IV   C) I, II, III, and IV   D) I, II, III, IV, and V

Comment: 74% of the students selected distractor (B), the correct answer. After displaying the histogram of responses, I discussed the process leading to the answer. Clicker use reminds us constantly that novices do not always make the connections that are obvious to the experts.

Class Discussion: A strong acid is *assumed* to be 100% ionized in dilute aqueous solution. This implies that HCl molecules are not present in 0.10 M HCl and the major species (present in relatively higher concentrations) in solution are H_3O^+, Cl^-, and H_2O.

Strong acid: HA(*aq*) + H_2O(*l*) → H_3O^+(*aq*) + A^-(*aq*) (Note the unidirectional arrow!)

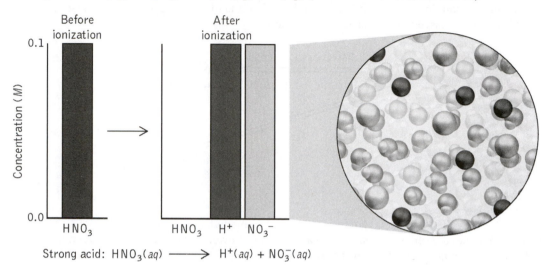

Strong acid: HNO_3(*aq*) ──────→ H^+(*aq*) + NO_3^-(*aq*)

Comment: The visual shown above presents both a column graph of relative amounts of the species in solution as well as the molecular/ionic scene. These different representations offer more than one way to think about the 0.10 M HCl solution (strong acid). Students will ask good questions about the relative numbers of water molecules and ions in the solution to get a better understanding of what we mean by "major" species. I have learned to anticipate these kinds of questions and enjoy working through the calculations ahead of lecture. We never stop learning! The next ConcepTest applies the Brönsted concept to the weak acid, CH_3COOH, used in the lecture demonstration.

ConcepTest
What major species are present in a 0.10 M CH_3COOH solution?

I) CH_3COOH II) H_2O III) H_3O^+ IV) CH_3COO^- V) OH^-

A) I and II   B) II, III, and IV   C) I, II, III, and IV   D) I, II, III, IV, and V

Comment: 9% selected distractor (A), the correct answer; 84% selected distractor (C). It is apparent that the students recognized the presence of undissociated CH_3COOH molecules, but they needed more information about the relative amounts of the ions and the molecules in solution. The discussion helps to address these essential concepts and make connections between pieces of information that many students have learned.

Class Discussion: A weak acid *is partially ionized* in solution. This implies that the majority of the species in solution (excluding water molecules) are molecules of HA, the weak acid; only a small number of the molecules have dissociated to form ions.

Weak acid: HA(*aq*) + H$_2$O(*l*) $\rightleftharpoons$ H$_3$O$^+$(*aq*) + A$^-$(*aq*)

(Note the reversible arrows used to represent the equilibrium process! The arrow pointing to the left is longer, confirming that equilibrium favors reactants; this is implied and is often not written this way in many textbooks as shown in the figure below.)

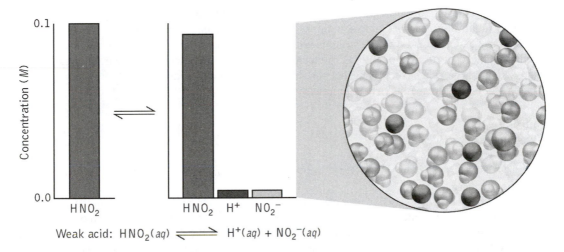

Weak acid: HNO$_2$(*aq*) $\rightleftharpoons$ H$^+$(*aq*) + NO$_2^-$(*aq*)

Comment: These visuals are extremely effective and it is a good idea to present the visuals for the strong and weak acids side by side to emphasize the similarities and differences between strong and weak acids. The chemical education literature has clearly documented the struggles of beginning students with the concept of "strong" and "weak" that is counterintuitive to the use of these terms in real life.

Next, I take the visualization to the next level by applying it to an acid-base neutralization reaction. Students learned about neutralization reactions in first-semester general chemistry and many of them would confidently recognize such a reaction and balance it. How well are they visualizing the processes represented by molecular, total ionic, and net ionic equations learned in first-semester general chemistry? The next ConcepTest addresses this question.

ConcepTest
What major species are present in solution when <u>equal volumes</u> of 0.10 M solutions of CH$_3$COOH and NaOH are reacted?

A) CH$_3$COOH, NaOH, and H$_2$O   B) H$_3$O$^+$, OH$^-$, and H$_2$O
C) NaC$_2$H$_3$O$_2$ and H$_2$O   D) Na$^+$, CH$_3$COO$^-$, and H$_2$O

Comment: 61% of the students responded correctly and 19% selected distractor (C). The latter group of students seemed quite distracted by the formula NaC$_2$H$_3$O$_2$ for sodium acetate and could not clearly distinguish between distractors (C) and (D). After they explained their rationale, I accepted (C) as an optional answer. In the original question, I had not included concentration and acknowledged my responsibility for the ensuing confusion. I explained that the formula $NaC_2H_3O_2$ implies solid that does not exist in dilute solutions of strong electrolytes; undissolved solid is present, for example, in a saturated solution.

Class Discussion: When equimolar amounts of a monoprotic acid (CH_3COOH) and a monoacidic base (NaOH) are present, the solution contains only salt and water at the neutralization point. In this example, the salt, $NaC_2H_3O_2$, is a strong electrolyte that is completely ionized (present as spectator ions Na^+ and CH_3COO^-) in aqueous solution.

Comment: This is a great opportunity to review, reinforce, and make connections between molecular, total ionic, and net ionic equations, and clarify the role of spectator ions in solution. Based on my experiences, students struggle with these concepts in first-semester general chemistry and resort to memorization. The lecture centered on the lecture demonstration and we discussed and reviewed several concepts pertaining to acids. I concluded the lecture with a ConcepTest that addresses a fairly common misconception.

ConcepTest
Which of these is a weak acid?

A) HBr(*aq*)   B) HCl(*aq*)   C) HF(*aq*)   D) HI(*aq*)

Class Discussion: HF is classified as a weak acid while HCl, HBr, and HI are considered strong acids. This result is not intuitive, based on the electronegativity of fluorine, which is the most electronegative element. A simple explanation is the strength of the relatively shorter HF bond compared to the longer HX bonds in the other hydrogen halides. Bond dissociation energy decreases down the group for the hydrogen halides. Another explanation for the relatively higher acid strength of HI compared to HF is the greater polarizability (and stability) of I^- compared to F^-.

Comment: Toward the end of the lecture, I inform students about the content of the next lecture to encourage them to read ahead and to prepare for ConcepTests related to the reading assignment. In the following lecture we covered quantitative aspects of pH and calculated the pH of 0.10 M HCl and 0.10 M CH_3COOH to test and confirm the validity of our qualitative predictions.

Many ideas for ConcepTests arise from questions that students ask in or outside class. It has been an exciting experience to learn from my students, who are often unbiased as scientists and ask very interesting questions. Students provide constructive remarks on course evaluations that help me grow as a reflective practitioner and appreciate the lifelong learning experiences of a science educator. Designing clicker questions that display visual information is a unique opportunity to combine art and science in creative ways.

CHAPTER 6   Administrative Issues

I hope that by now you are excited about the teaching and learning possibilities that clickers and CRS systems can create for you and your students. But before you pose your first clicker question to students, there are some behind-the-scenes administrative issues that require your consideration. Thinking through and documenting these administrative issues now will allow you to spend valuable time later focusing on more effective interactions, more engaging and challenging activities, and further improvements in long-term understanding and retention. In the hopes that I can make your implementation of clickers as stress-free as possible, in this chapter I share with you my thoughts on documenting clicker use in your course syllabus; grading; cheating; dealing with absences, malfunctioning clickers, and other excuses; and posting clicker questions online.

Documenting Clicker Use in Your Course Syllabus

I have found that to facilitate student acceptance of clicker use, it is very important to document their use and the class policies that are affected by their use. This includes explaining the rationale for using clickers in your class, how student clicker transmitters will be registered, how class participation using clickers is assessed and rewarded, and what your class policies will be for absences, malfunctioning clickers, and cheating. Having this information clearly documented in your course syllabus makes transparent the expectations of student behavior, and saves you from later dealing with those that would seek to take advantage of poorly documented policies. Since most first-year students will be new to using clickers and CRS systems, they will appreciate knowing how the clicker-use affects their grades and the additional cost that may be incurred.

Here is the statement from my Fall 2008 second-semester general chemistry syllabus that documents clicker use in the course:

> **CONCEPTESTS AND iCLICKERS:** Each student must purchase an iClicker to participate in ConcepTests presented in lecture to actively engage all students in the learning process. There will be three to five ConcepTest questions in each lecture beginning with the second lecture on Wednesday, August 27. A ConcepTest question is usually presented in multiple-choice format. The student typically has 1 to 1.5 minutes to respond using the iClicker; peer collaboration is often encouraged. A histogram of responses is displayed at the end of each exercise. Misconceptions will be addressed, and the processing of information will be discussed as necessary.
>
> A maximum of 60 *grade points* can be earned from participation in ConcepTests. A correct answer receives 2 *clicker points* and an incorrect answer receives 1 *clicker point* (for participation). At the end of the semester, a student's clicker points are added and then increased by 10% to allow for absences, malfunctioning clickers, and all other possible problems. These clicker points are then normalized to the maximum of 60 grade points. For example, if there are 100 ConcepTests, there are 200 possible clicker points. A student who earns 150 clicker points (after the 10% increase) will receive 45 grade points.

To Grade or Not to Grade

When my colleague and I first used clickers in August 2003, in a class of more than 950 students, we decided to offer bonus points in lieu of making clicker participation a mandatory part of the course grade because there were a few technology-related problems that went along with the early infrared (IR) systems. Students remarked that they appreciated this approach because then they did not feel pressured to get the correct answer every time. However, our data showed that this lax grading policy gave the impression that clicker participation was not important and consequently many of our best students did not participate in class; some did not even bother to purchase the CRS transmitters.

Because students only receive the full benefit of clickers when the majority of their classmates participate, I have since made clicker participation a mandatory, graded part of the course grade, and I suggest that you do the same. In my class (course syllabus documented above), I assign points to clicker use, which count toward 5% of a student's course grade. My data showed a significant increase in class attendance, from 55 to 60% prior to clicker use, to around 85% after clickers contributed to the course grade.

Cheating

Of course, if something is graded then someone is likely to try to cheat. I am happy to report that, at least in my classes thus far, cheating has not been a serious issue, and the honor system usually works quite well. The only incident of cheating that I can recall is a student who volunteered to click in responses for a classmate who felt quite ill upon entering the classroom and had to leave. The student was reprimanded for poor judgment and warned that she/he would receive a failing grade for the course if caught a second time. As a deterrent, my syllabus now contains a clause stating that "a student caught cheating with clickers will receive an F [failing grade] for the semester," and my grading of clicker participation is such that an absence will not create a similar situation again.

Dealing with Absences, Malfunctioning Clickers, and Other Excuses

Speaking of absences, because I grade clicker participation—and strongly suggest you do too—I inevitably hear from students who have missed class, legitimately or otherwise, and want to ameliorate the negative effect on their grade this absence will entail. In addition to missing class, I will often hear other excuses from students as to why they could not participate in the clickers for that class. Some of the most common excuses are "I did not bring my clicker today," "my battery just died," and "my clicker is not working properly."

I have learned to deal with a variety of such problems by either dropping a certain number of questions at the end of the semester, or by increasing each score by 5 to 10% prior to normalizing clicker points. This policy is always documented in my course syllabus. Be warned that a few zealous students will still come to you with these excuses hoping for an option!

Posting Clicker Questions Online

I have often been asked by students to post online the clicker questions from the previous lecture. In addition, many students request that I post answers separately from the questions. I have often agreed to do so, since I emphasize conceptual understanding and believe that posting the questions and answers (highlighted answers) can be beneficial to beginning students who are accustomed to algorithmic approaches to learning. My goal in using clickers has always been to encourage and support my students, and I don't consider posting the clicker questions online that different from posting lecture notes or presentations online in support of students' learning.

Conclusion

I hope this part of the handbook, "Part I—Clickers in the Classroom," has convinced you of the transformative possibilities that CRS systems create and that you are now eager to start using clickers in your own course. This handbook contains additional resources to assist you with the implementation of clickers in your courses.

Part II of this handbook, "Questions for General Chemistry," contains class-tested clicker questions on topics covered by most first- and second-semester general chemistry courses, and a little over half of them include histograms documenting actual student performance. The accompanying CD to this book contains these ready-made questions in PowerPoint format so that you may modify them if you wish, and easily import them into the CRS system of your choosing. In Part III, "Lecture Demonstrations," I offer suggestions for how clicker use can be extended beyond testing student comprehension of conceptual material by testing student understanding of what they witness in chemistry demonstrations of the type readily performed in lecture.

As you use and modify the resources contained in this handbook, I encourage you to join and participate in the growing community of chemistry educators that are using clickers and sharing their experiences. Whether it's an online chemistry forum or a conference panel, you will find supportive instructors enthusiastic about improving the teaching and learning in their classes through clickers.

CHAPTER 7   The Study of Matter and Measurement

1. Which of these atomic and/or molecular views represent pure substances?

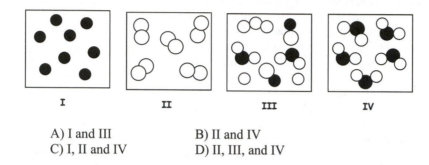

| I | II | III | IV |

    A) I and III   B) II and IV
    C) I, II and IV   D) II, III, and IV

Correct Answer: C
Level of Difficulty: Easy/Moderate
Keyword(s): Atomic/molecular views; pure elements and compounds, atoms and molecules
Histogram:

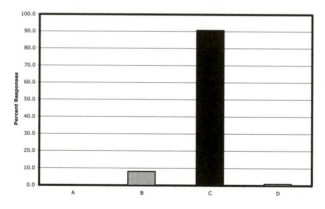

Comments: Visualization skills should be emphasized throughout the course and tested on homework and exams. We have found that many freshmen lack these skills. When students become comfortable with atomic and molecular views, questions like this one become easy. The histogram shows that 91% answered correctly. Peer collaboration was encouraged when the clicker question was presented. This question addresses the misconception that compounds (as in IV) are not pure substances because they contain two or more different kinds of atoms in the unit particle. We highly recommend discussion of the thought process used to arrive at the correct answer; it is very important to address why other answers are incorrect. The instructor may emphasize that elements such as metals and the inert gases are monoatomic at the particulate level, elements such as H_2, N_2, O_2, F_2, Cl_2, Br_2, and I_2 are diatomic, and many compounds and elements such as P_4 and S_8 are polyatomic. Reinforcement of the diatomic elements is essential as many students forget this information when writing equations, dealing with gas law problems, or working with the Born-Haber cycle.

2. Which equation is the correct symbolic representation for the sublimation of iodine crystals?

A) $I_2(g) \rightarrow I_2(s)$ B) $I_2(s) \rightarrow I_2(g)$
C) $I_2(s) \rightarrow I_2(l)$ D) $I_2(l) \rightarrow I_2(s)$

Correct Answer: B
Level of Difficulty: Moderate
Keyword(s): Physical changes
Histogram:

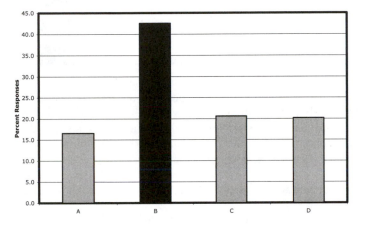

Comments: This question could accompany a lecture demonstration of the sublimation of iodine crystals (details of such a setup are included in Part III of this handbook). As the histogram indicates, only 43% of the students answered correctly and significant numbers of students selected each of the other distractors. Many freshmen are not very confident using scientific terms such as condensation, deposition, and sublimation that refer to specific physical changes. Another challenge is the correct interpretation of symbolic representations. We often misjudge prior knowledge of our students; it is more effective to review and reinforce these very basic concepts.

3. Dry ice sublimes at 25°C and 1 atmosphere pressure. Which equation is the correct symbolic representation for the change that occurs?

A) $CO_2(g) \rightarrow C(s) + O_2(g)$ B) $CO_2(g) \rightarrow CO_2(s)$
C) $CO_2(s) \rightarrow CO_2(l)$ D) $CO_2(s) \rightarrow CO_2(g)$

Correct Answer: D
Level of Difficulty: Easy
Keyword(s): Physical changes

Note: Histogram and comments for this question appear on the next page.

Histogram:

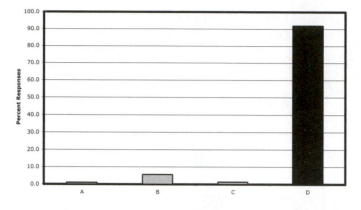

Comments: This question is an alternative to the previous one and could accompany a lecture demonstration of the sublimation of dry ice (details of such a setup are included in Part III of this handbook). Students are generally more familiar with dry ice; the histogram shows that 92% of the students selected the correct answer. This question addresses symbolic representations and the importance of physical states in thermochemical equations.

4. Which of these are chemical properties of matter?

   I) Corrosiveness   II) Density   III) Flammability   IV) Melting point

   A) I and II   B) I and III   C) II and IV   D) III and IV

Correct Answer: B
Level of Difficulty: Easy
Keyword(s): Chemical properties
Comments: We should provide examples of chemical processes that describe corrosiveness and flammability. For example, many acids react with metals causing them to corrode. Flammability results from combustion reactions (burning in oxygen).

5. Which equation best represents the change that takes place when water is electrolyzed?

   A) $H_2O(l) \rightarrow H_2O(g)$ B) $H_2O(g) \rightarrow H_2O(l)$
   C) $2\ H_2O(l) \rightarrow 2\ H_2(g) + O_2(g)$ D) $2\ H_2(g) + O_2(g) \rightarrow 2\ H_2O(l)$

Correct Answer: C
Level of Difficulty: Easy/Moderate
Keyword(s): Chemical changes

Note: Histogram and comments for this question appear on the next page.

Histogram:

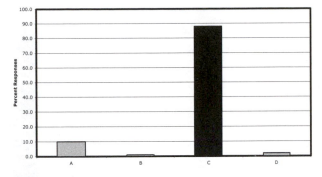

Comments: This question helps to test prior knowledge and to assess the ability of students to integrate concepts with equations and symbolic representations. The histogram confirms that the majority (88%) of the students are familiar with the electrolysis of water. We encourage you to perform a lecture demonstration (details of such a setup are included in Part III of this handbook) and use atomic and molecular graphics to visually depict the chemical change. This is also a great example to engage students about the challenges to producing hydrogen gas as a clean-burning fuel starting with water.

6. Extensive properties of a pure substance depend on sample size whereas intensive properties are characteristic of that substance. Which of these properties are intensive?

I) Color   II) Mass   III) Density

A) I and II   B) I and III   C) II and III   D) I, II and III

Correct Answer: B
Level of Difficulty: Moderate
Keyword(s): Extensive/Intensive properties
Comments: Our students struggle with the distinction between extensive and intensive properties. A useful hint is that an extensive property depends on the "extent" or amount (sample size) of the substance. We could provide additional examples of intensive and extensive properties.

7. Identify the best match between the dimension or quantity and its correct SI base unit.

	Dimension or Quantity	Unit
A)	Mass	Gram
B)	Length	Kilometer
C)	Time	Minute
D)	Temperature	Celsius
E)	Amount of substance	Mole

Correct Answer: E
Level of Difficulty: Easy/Moderate
Keyword(s): SI units

Histogram:

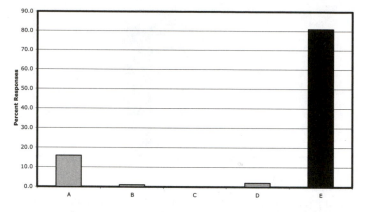

Comments: The histogram shows that 81% of the students selected the correct answer. About 17% picked distractor (A); the correct SI unit of mass is the kilogram. It is recommended that instructors use this question to test prior knowledge and use this opportunity to review the correct SI base unit for each quantity or dimension listed. This may be preferred over a list of all the SI base units.

8. Select the correct relationship between these metric units of length or distance.

A) 1 km = 100 m

B) 1 mm = 10 $\overline{\text{cm}}$

C) 1 nm = 10^9 m

D) 10^6 μm = 1 m

Correct Answer: D
Level of Difficulty: Easy/Moderate
Keyword(s): Metric prefixes and unit factors
Comments: Students should learn to recognize some of the metric prefixes and to distinguish between large and small numbers. For example, they must know that a nanometer (nm) is smaller than a meter; if the prefix nano- means 10^{-9}, they must confidently recognize that 10^9 nm = 1 m. Additional visuals can be developed to understand prefixes and relationships.

9. In which of these measured values are the zeros not significant figures?

I) 0.0591 cm   II) 504 g   III) 2.70 m   IV) 5300 L

A) I and II

B) II and III

C) I and IV

D) I, III, and IV

E) II, III, and IV

Correct Answer: C
Level of Difficulty: Moderate
Keyword(s): Significant figures

Histogram:

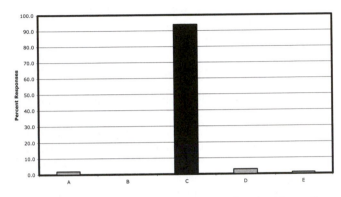

Comments: This is an excellent question to review the rules for significant figures because each of the choices I–IV illustrates the role of the zero and why it is/is not significant. The histogram shows excellent performance (94% correct) because the rules were discussed prior to presenting the question. The question would be ideal to assess student accountability for reading-assigned sections prior to class.

10. A metal sample is hammered into a rectangular sheet with an area of 31.2 ft^2 and an average thickness of 2.30×10^{-6} cm. If the mass of this sample is 0.4767 g, predict the identity of the metal. The density of the metal is shown in parenthesis. Useful information: 1 in = 2.54 cm

A) Aluminum (2.70 g/cm^3)   B) Copper (8.95 g/cm^3)
C) Gold (19.3 g/cm^3)   D) Zinc (7.15 g/cm^3)

Correct Answer: D
Level of Difficulty: Challenging
Keyword(s): Density; dimensional analysis
Histogram:

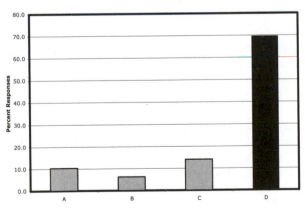

Comments: This is a challenging question. Students should recognize the relationship between mass, volume, and density. This section addresses a common math error where students forget to square both the number and the unit. This question appeared on the first hour exam and only 69% answered correctly. It is very important to include clicker questions with the same level of difficulty as exam questions. Students often complain that we use relatively easy examples in class and present challenging questions on exams.

11. Select the correct statement pertaining to energy, relative energies, and stability.

  A) In nature, systems of higher energy are typically more stable than those of lower energy.
  B) The chemical potential energy of a substance results from the relative positions and the attractions and repulsions among all its particles.
  C) Kinetic energy is converted to potential energy as a boulder rolls downhill.
  D) A system of two balls attached by a spring is relatively more stable when the spring is stretched rather than relaxed.

Correct Answer: B
Level of Difficulty: Moderate
Keyword(s): Potential/kinetic energy; energy and relative stabilities
Histogram:

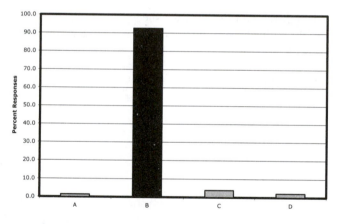

Comments: Our students are introduced to the concept of potential versus kinetic energy in middle and high school. We present the relationship between stability and relative energies early in the course and reinforce them throughout. This concept lays the foundation for understanding and interpreting energy level diagrams. The histogram shows that the majority (93%) of the students answered correctly.

12. Neon has a boiling point of 27 K. Express this temperature in degrees Fahrenheit.

  A) 352°F   B) 168°F   C) −246°F   D) −411°F

Correct Answer: D
Level of Difficulty: Moderate/Challenging
Keyword(s): Temperature scales and conversions between them

Histogram:

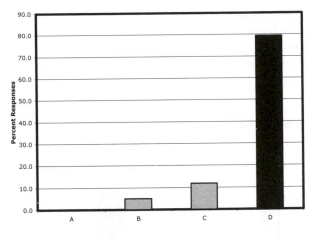

Comments: This question requires conversion from degrees Kelvin to °C to °F. Many students are generally comfortable with the first conversion; however, they struggle with using the equation to convert °C to °F. About 79% answered correctly on the first hour exam. We encourage you to set up the problem and present the complete solution explaining the mathematical operations as needed.

13. The melting point of pure benzoic acid is 122°C. Data obtained by four students in a laboratory experiment are shown. Which student's data are precise but not accurate?

Student A	Student B	Student C	Student D
115°C	119°C	122°C	118°C
112°C	118°C	121°C	120°C
118°C	119°C	122°C	124°C
116°C	120°C	121°C	126°C

A) Student A   B) Student B   C) Student C   D) Student D

Correct Answer: B
Level of Difficulty: Moderate
Keyword(s): Precision and accuracy
Comments: This question allows students to apply their understanding of precision and accuracy. Student C's data are both precise and accurate while Student B's data are precise but not accurate. Student A's data are neither precise nor accurate. Student D's data are not precise and the average is the expected result.

14. Select the correct statement about the modern periodic table.

        A) Tin is a transition element.
        B) Lead is a nonmetal.
        C) Antimony is a metalloid.
        D) Elements are arranged in order of increasing atomic mass.

Correct Answer: C
Level of Difficulty: Moderate
Keyword(s): Atomic symbols and periodic table
Histogram:

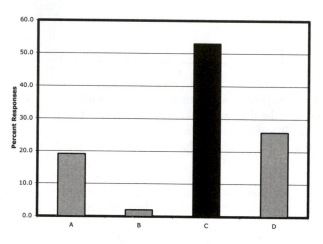

Comments: This question attempts to present chemistry in the context of the periodic table early in the course to assess prior knowledge. Only 53% of the students selected the correct answer confirming the need for review and reinforcement throughout the course.

CHAPTER 8   Introduction to Atomic Structure

1. Dalton's atomic theory (1808) has 4 postulates. Which postulate remains true today?

A) All matter consists of atoms, which are indivisible particles of an element.
B) Atoms of one element cannot be converted into atoms of another element.
C) Atoms of an element are identical in mass and other properties.
D) Compounds are formed by the chemical combination of two or more elements in specific ratios.
E) None of the above remains true today.

Correct Answer: D
Level of Difficulty: Easy/Moderate
Keyword(s): Postulates of Dalton's atomic theory
Histogram:

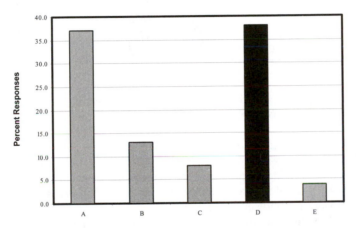

Comments: These results were surprising; 37% selected response (A) compared to 38% who selected the correct answer. This question was designed to review the reading assignment as well as to test understanding and retention from high school chemistry. We can facilitate learning and enhance retention by reviewing and reinforcing the material when opportunities arise.

2. Carefully consider these four postulates of Dalton's atomic theory:

I) All matter consists of atoms, which are indivisible particles of an element.
II) Atoms of one element cannot be converted into atoms of another element.
III) Atoms of an element are identical in mass and other properties.
IV) Compounds are formed by the chemical combination of two or more elements in specific ratios.

Which postulate is no longer valid after the discovery of nuclear reactions?
(e.g., ^{226}Ra → ^{222}Rn + α-particle)

A) I   B) II   C) III   D) IV

Correct Answer: B
Level of Difficulty: Easy/Moderate
Keyword(s): Postulates of Dalton's atomic theory
Histogram:

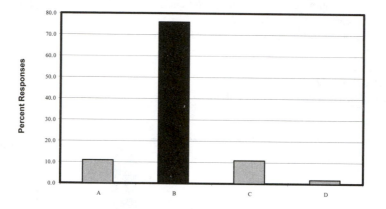

Comments: The example provided in the question may have helped to see patterns and make connections. Students did reasonably well on this question.

3. Which statement is <u>not true</u> concerning cathode rays?

        A) They originate from the negative electrode in a cathode ray tube.
        B) They travel in straight lines in the absence of electric or magnetic fields.
        C) They are made up of electrons.
        D) The properties of cathode rays are dependent on the cathode material.

Correct Answer: D
Level of Difficulty: Easy/Moderate
Keyword(s): Properties of cathode rays
Comments: The use of visuals or a lecture demonstration can be very effective in covering this topic.

4. Thomson obtained a value of -1.759×10^{11} C/kg (Coulombs/kilogram) for the charge-to-mass (e/m) ratio for the electron. Millikan determined the charge (e) on the electron to be -1.602×10^{-19} C. Calculate the mass of the electron.

        A) 1.098×10^{30} kg   B) 9.107×10^{-31} kg
        C) 2.817×10^{-8} kg   D) none of these

Correct Answer: B

Level of Difficulty: Easy/Moderate
Keyword(s): Charge-to-mass ratio, dimensional analysis; ball park estimates
Histogram:

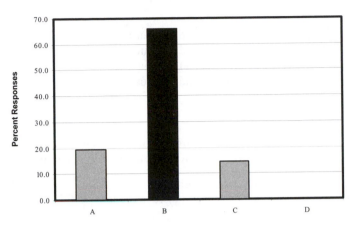

Comments: This problem could have been solved using dimensional analysis. Some students struggle with dimensional analysis; required use in class, in collaboration with other students, may be helpful.

5. Which of these species has the highest number of electrons?

A) $_{20}Ca$ B) $_{19}K^+$ C) $_{16}S^{2-}$ D) $_{15}P^{3-}$

Correct Answer: A
Level of Difficulty: Easy
Keyword(s): Number of electrons in neutral and charged species
Histogram:

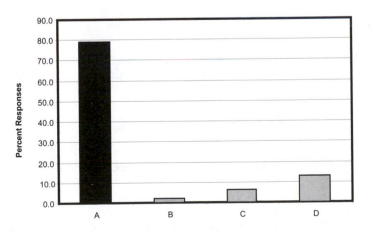

Comments: About 79% of the students selected the correct answer. This is a good question to compare neutral atoms and ions in regard to the number of electrons in each species.

6. The ion $^{45}Sc^{3+}$ has[3]

        A) 24 electrons, 21 protons, and 24 neutrons
        B) 18 electrons, 21 protons, and 24 neutrons
        C) 24 electrons, 24 protons, and 21 neutrons
        D) 18 electrons, 24 protons, and 21 neutrons

Correct Answer: B
Level of Difficulty: Easy/Moderate
Keyword(s): Mass number/atomic number relationships in positively charged ions
Comments: This is a fairly easy question and provides drill and practice in the classroom.

7. How many electrons, protons, and neutrons are there in the ion $^{77}_{35}X^{3-}$?

        A) 35 electrons, 35 protons, and 35 neutrons
        B) 35 electrons, 38 protons, and 35 neutrons
        C) 38 electrons, 35 protons, and 42 neutrons
        D) 38 electrons, 38 protons, and 42 neutrons

Correct Answer: C
Level of Difficulty: Easy/Moderate
Keyword(s): Mass number/atomic number relationships in negatively charged ions
Comments: The use of X rather than a recognizable symbol can be a useful learning approach.

8) The ion X^{2+} has 24 electrons. A possible identity for X is

    A) Cr   B) Ti   C) Fe   D) Co

Correct Answer: C
Level of Difficulty: Moderate
Keyword(s): Mass number/atomic number relationships in positively charged ions
Comments: This is a higher-level question that does not immediately establish the relationship between the symbol and the number of protons.

9. $^{35}Cl^-$, ^{40}Ar, and $^{39}K^+$ all have the same

        A) atomic number.   B) mass number.
        C) number of electrons.   D) number of neutrons.

Correct Answer: C
Level of Difficulty: Easy/Moderate
Keyword(s): Mass number/atomic number relationships in neutral and charged species

Comments: We use this question to integrate several concepts and to emphasize the differences between atoms and ions.

10. Two stable isotopes of an element have isotopic masses of 10.0129 amu and 11.0093 amu. The atomic mass is 10.81. Which isotope is more abundant?

> A) There is insufficient information to answer the question.
> B) There are equal amounts of each isotope.
> C) The isotope with a mass of 10.0129 amu is more abundant.
> D) The isotope with a mass of 11.0093 amu is more abundant.

Correct Answer: D
Level of Difficulty: Easy/Moderate
Keyword(s): Isotopes and percent abundance, qualitative predictions
Histogram:

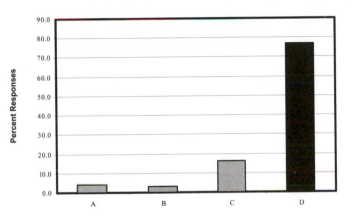

Comments: This question helps students to gain confidence in making qualitative predictions rather than plugging into an equation.

11. Which one of the four postulates of Dalton's atomic theory (1808) was disproved when isotopes were discovered?

> A) All matter consists of atoms, which are indivisible particles of an element.
> B) Atoms of one element cannot be converted into atoms of another element.
> C) Atoms of an element are identical in mass and other properties.
> D) Compounds are formed by the chemical combination of two or more elements in specific ratios.

Correct Answer: C
Level of Difficulty: Easy
Keyword(s): Dalton's atomic theory

Histogram:

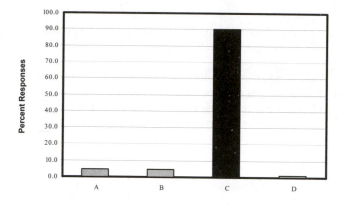

Comments: This question reinforces Dalton's atomic theory after a discussion of isotopes. Student performance confirms a good grasp of the concepts.

12. The two major isotopes of bromine are ^{79}Br and ^{81}Br. Assume that the masses of the ^{79}Br and ^{81}Br isotopes are 79.00 and 81.00 amu, respectively. The weighted average atomic mass of bromine is 79.90 amu. What are the relative % abundances of each isotope? Estimate WITHOUT detailed calculations!

	% Abundance of ^{79}Br	% Abundance of ^{81}Br
A)	79.0%	21.0%
B)	19.0%	81.0%
C)	35.1%	64.9%
D)	55.0%	45.0%

Correct Answer: D
Level of Difficulty: Moderate
Keyword(s): Isotopes and percent abundance, qualitative predictions

Note: Histogram and comments for this question appear on the next page.

Histogram:

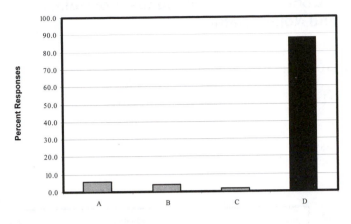

Comments: Encouraging students to make qualitative predictions using numerical data can be a learning experience that saves them time on tests. Many students feel more comfortable using the equation.

CHAPTER 9 Introduction to Compounds, Formulas, and Nomenclature

1. Select the transition element.

         A) Lead   B) Potassium   C) Silver   D) Tin

Correct Answer: C
Level of Difficulty: Easy
Keyword(s): Periodic table basics
Histogram:

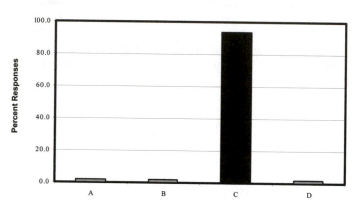

Comments: The effectiveness of the periodic table as a valuable resource is emphasized throughout the course. Students are encouraged to master symbols, names, and classification of elements by groups and periods.

2. Select the correct statement about the modern periodic table.

         A) Beryllium is an alkali metal.
         B) Bismuth is a transition metal.
         C) Sulfur is a halogen.
         D) Silicon is a metalloid.

Correct Answer: D
Level of Difficulty: Easy/Moderate
Keyword(s): Periodic table basics

Histogram:

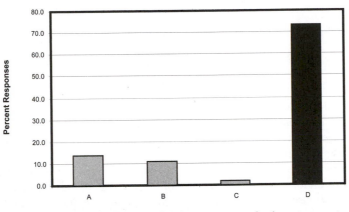

Comments: This question was included on the final exam to test for long-term retention of knowledge and we had hoped for a better performance.

3. Classify each statement about the periodic table as true (T) or false (F). Then, select the answer with the three correct letters (example, TTT).

I) In the modern periodic table, elements are arranged in order of increasing number of protons.
II) Lead is a transition element.
III) Radon is an inert gas.

A) FTF   B) TFT   C) FFT   D) TTF

Correct Answer: B
Level of Difficulty: Easy/Moderate
Keyword(s): Periodic table basics
Histogram:

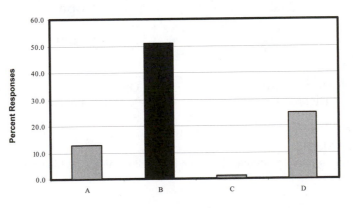

Comments: These results confirm that students are often overwhelmed by the information overload. We continue to look for opportunities to reinforce concepts to help students in the learning process.

4. Predict the formula of the binary ionic compound formed by magnesium and phosphorus.

  A) Mg_2P B) MgP_2 C) Mg_3P_2 D) Mg_2P_3

Correct Answer: C
Level of Difficulty: Moderate
Keyword(s): Binary compound between metal and nonmetal
Comments: We review Lewis structures early in the course to reinforce understanding of the charges on the ions and to explain the ratios in which the ions combine to form the neutral compound. Students tend to cling to the tricks learned in high school to write correct formulas by memorizing charges on ions. The use of the periodic table to obtain this information is strongly emphasized. We have observed improvements in learning and retention as a result of this teaching approach to binary ionic compounds. About 93% of the students answered this question correctly.

5. Predict the formula for the binary ionic compound formed by aluminum and oxygen.

  A) Al_2O_3 B) Al_3O_2 C) Al_2O D) AlO_2

Correct Answer: A
Level of Difficulty: Moderate
Keyword(s): Binary compound between metal and nonmetal
Comments: This question may be used to test understanding and retention from high school chemistry. About 87% of the students answered correctly. Even though a significant number of students responded correctly, we reviewed the process leading to the selection of the correct answer.

6. Predict the correct name of the compound represented in the box.

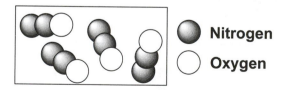

  A) Nitrogen oxide   B) Oxygen nitride
  C) Dinitrogen monoxide D) Nitrogen dioxide

Correct Answer: C
Level of Difficulty: Moderate
Keyword(s): Naming binary compounds between nonmetals; picturing binary molecules

Histogram:

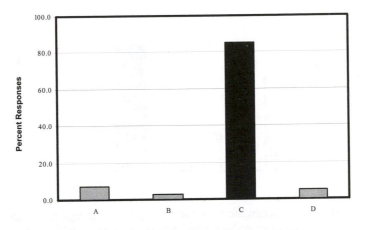

Comments: Picturing molecules makes chemistry interesting. Integration of molecular views and naming compounds or writing formulas will help students to develop and reinforce visualization skills. The shapes of molecules intrigue some students. Students should recognize the binary compound between two nonmetals and deduce the correct formula; about 15% of the students did not arrive at the correct answer. This is a good example to remind students that prefixes are used only in naming binary compounds between nonmetals.

7. Which name on the right correctly corresponds to the formula shown on the left?

	Correct Formula	Name
A)	$HBr(aq)$	Bromic acid
B)	$AlCl_3$	Aluminum trichloride
C)	BF_3	Boron trifluoride
D)	Cu_2S	Copper sulfide

Correct Answer: C
Level of Difficulty: Moderate/Challenging
Keyword(s): Naming binary compounds

Note: Histogram and comments for this question appear on the next page.

Histogram:

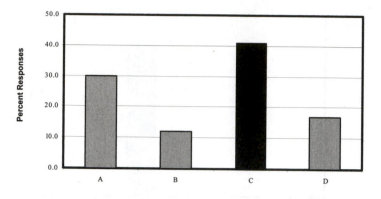

Comments: This question summarizes the rules for naming binary inorganic compounds, and the distribution of results shows that the students need more time to practice the rules and apply them correctly. We have observed similar results in class for several semesters when identical questions were presented.

8. Select the correct statement for naming binary inorganic compounds.

  A) K_2O is dipotassium oxide.
  B) $H_2Se(aq)$ is hydroselenic acid.
  C) CuO is copper oxide.
  D) $MgCl_2$ is magnesium(II) chloride.

Correct Answer: B
Level of Difficulty: Moderate
Keyword(s): Review of rules for naming compounds
Comments: This question also summarizes the rules for naming binary inorganic compounds. About 83% of the students answered correctly.

9. Which compound is named correctly? Assume the formulas are correct.

  A) Na_3P; sodium phosphate
  B) $CaBr_2$; calcium dibromide
  C) Cu_2O; copper(II) oxide
  D) $Fe_2(SO_4)_3$; iron(III) sulfate

Correct Answer: D
Level of Difficulty: Moderate
Keyword(s): Connecting names and formulas of ionic compounds; polyatomic ions

Histogram:

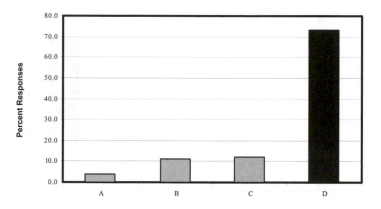

Comments: Students tend to struggle with learning formulas and charges of polyatomic ions. This question was presented on the first hour exam about 4 weeks into the semester. It is a good idea to reinforce this concept frequently prior to writing balanced equations.

10. Which compound is represented by the correct formula? Assume the names are correct.

        A) Magnesium phosphide; Mg_2P_3
        B) Magnesium phosphate; $Mg_3(PO_4)_2$
        C) Magnesium hydrogen phosphate; $Mg(HPO_4)_2$
        D) Magnesium dihydrogen phosphate; MgH_2PO_4

Correct Answer: B
Level of Difficulty: Moderate
Keyword(s): Connecting names and formulas of ionic compounds; polyatomic ions
Comments: This question was presented on the first hour exam. Students need a lot of drill and practice with polyatomic ions; 80% answered correctly.

11. Select the compound that contains both ionic and covalent bonds.

        A) CaO   B) CO_2 C) $CaCl_2$ D) $CaCO_3$

Correct Answer: D
Level of Difficulty: Easy/Moderate
Keyword(s): Ionic and covalent bonding

Note: Histogram and comments for this question appear on the next page.

Histogram:

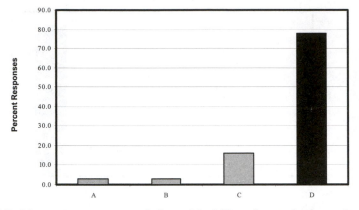

Comments: We find it very necessary to reinforce this skill so that students continue to connect formulas with bonding concepts. A significant 16% of the students selected distractor (C) ($CaCl_2$) compared to 78% who answered correctly. Our experience on several occasions has shown that students mistake Cl_2 in the formula for the diatomic chlorine molecule that is covalent. This was confirmed when students drew pictures during recitation of the ions in aqueous $CaCl_2$ and included Cl_2^{2-} ions!

12. Which compound is represented by the correct formula? Assume the names are correct.

A) Iron(III) nitrate; $Fe(NO_3)_3$ B) Calcium hydroxide; $CaOH$
C) Lithium nitride; LiN D) Potassium sulfate; KSO_4

Correct Answer: A
Level of Difficulty: Moderate/Challenging
Keyword(s): Matching names and formulas of inorganic compounds; using polyatomic ions
Comments: This question was presented on the course final exam to test long-term retention and 80% answered correctly. This appears to be consistent with the grades that we assign; about 20% of our students earn D's and F's in the course. Over the years we have learned that 15–20% of the students who enroll in first-semester general chemistry have never taken a high school chemistry course.

13. Which formula containing polyatomic ions is correct?

A) $MgNO_3$ B) NH_4CO_3 C) $Na(PO_4)_3$ D) $Al_2(SO_4)_3$

Correct Answer: D
Level of Difficulty: Moderate/Challenging
Keyword(s): Compounds containing polyatomic ions
Comments: About 98% of the students selected the correct answer in class after collaboration with peers. Some of the polyatomic ion formulas were presented in lecture preceding the question; 98% of the students answered correctly confirming some short-term mastery of the logic behind writing correct formulas.

CHAPTER 10 Mole Concept and Stoichiometry

1. Select the correct statement pertaining to the mole concept.

        A) A mole of oxygen gas contains 6.022×10^{23} atoms.
        B) A mole of chlorine gas contains $2 \times 6.022 \times 10^{23}$ chlorine molecules.
        C) A mole of CO_2 contains 2 moles of oxygen atoms.
        D) A mole of ammonia has a mass of 34.0 g.

Correct Answer: C
Level of Difficulty: Moderate/Challenging
Keyword(s): Mole concept and numbers of unit particles/atoms; mass of a mole
Histogram:

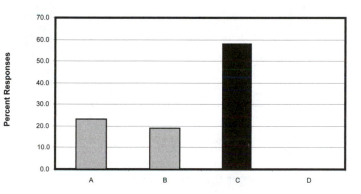

Comments: The distribution of responses confirms that students are challenged by the mole concept especially as it pertains to the number of unit particles. Our students appreciate some of the cartoons that we draw to make these connections.

2. What is the total number of atoms in 1 mole of aluminum iodide?

        A) 2   B) $2 \times 6.022 \times 10^{23}$
        C) $3 \times 6.022 \times 10^{23}$ D) $4 \times 6.022 \times 10^{23}$

Correct Answer: D
Level of Difficulty: Moderate
Keyword(s): Mole, formula units of ionic compound, and atoms
Comments: Only 75% of the students responded correctly in class after peer collaboration.
Visualizing the unit particles (atoms, molecules, formula units) and atoms in a mole is a challenge.

3. How many hydrogen atoms are there in 34 g of ammonia, NH_3? The molar mass of NH_3 is 17.0 g/mol.

A) 6
B) $2 \times 6.022 \times 10^{23}$
C) $3 \times 6.022 \times 10^{23}$
D) $6 \times 6.022 \times 10^{23}$

Correct Answer: D
Level of Difficulty: Moderate
Keyword(s): Mass of a mole and number of unit particles/atoms
Histogram:

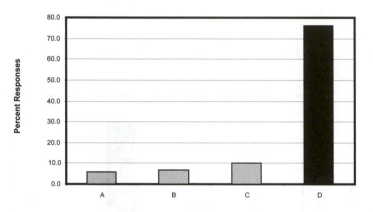

Comments: About 24% of the students did not obtain the correct answer. The question may appear relatively simple, especially to an experienced instructor; we need to address some of the difficulties that our students face prior to mastering the mole concept.

4. How many oxygen atoms are there in 17.1 g of sucrose, $C_{12}H_{22}O_{11}$? Molar mass = 342.0 g/mol

A) 5.85×10^{21}
B) 3.01×10^{22}
C) 3.31×10^{23}
D) 7.11×10^{24}

Correct Answer: C
Level of Difficulty: Moderate/Challenging
Keyword(s): Moles, mass, molecules and atoms in a sample

Note: Histogram and comments for this question appear on the next page.

Histogram:

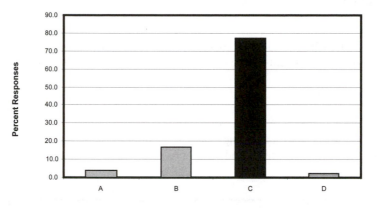

Comments: This question was presented on the first hour exam and 77% answered correctly; we were pleased to see increased confidence in using the mole concept. We would use this question in a lecture setting in the future. Most of the ConcepTests focus on conceptual understanding; problem-solving questions link conceptual understanding and quantitative/mathematical skills and should be included when possible.

5. Ascorbic acid, or vitamin C ($C_6H_8O_6$), an essential vitamin, cannot be stored by the body and must be present in the diet. If a typical tablet contains 500.0 mg of vitamin C, how many carbon atoms are present in the tablet? The molar mass of vitamin C is 176.1 g/mol.

A) 1.026×10^{22} carbon atoms   B) 9.051×10^{21} carbon atoms
C) 3.522×10^{21} carbon atoms   D) 7.841×10^{20} carbon atoms

Correct Answer: A
Level of Difficulty: Moderate
Keyword(s): Moles, mass, molecules and atoms in a sample
Comments: This question was also presented on the first hour exam and students appear to have mastered the mole concept. We would consider using this in lecture or recitation if time permits.

6. Examine the condensed structural formulas shown below:

I) Acetic acid (main ingredient in vinegar), CH_3COOH
II) Formaldehyde (used to preserve biological specimens), HCHO
III) Ethanol (alcohol in beer and wine), CH_3CH_2OH

For which molecule(s) are the empirical formula(s) and the molecular formula(s) the same?

A) II   B) I and II   C) II and III   D) I, II, and III

Correct Answer: C
Level of Difficulty: Easy/Moderate
Keyword(s): Structural formulas, empirical and molecular formulas

Comments: Only 76% of the students grasped the concept after presentation in lecture. Drill and practice using electronic homework and recitation tutorials were used to facilitate mastery of this concept.

7. Ammonium nitrate, when heated, decomposes into nitrogen gas, oxygen gas, and water vapor. What is the sum of the coefficients in the balanced equation using smallest integer coefficients?

A) 3   B) 5   C) 7   D) 9

Correct Answer: D
Level of Difficulty: Moderate/Challenging
Keyword(s): Writing correct formulas and balancing equations
Histogram:

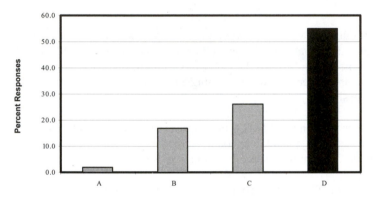

Comments: Only 55% answered correctly. The distribution of responses may be related to errors in writing correct formulas. Our experience has shown that students frequently forget the diatomic molecules. Perhaps, some students struggled with the polyatomic ions in ammonium nitrate. Our strategy in combining several concepts is to emphasize the importance of these essential skills and to encourage students to master basic skills for success in chemistry.

8. Consider the molecular view of reactants converted to a product in the boxes shown below:

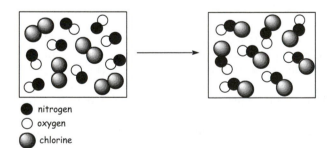

● nitrogen
○ oxygen
◐ chlorine

Which balanced equation best represents this reaction?

A) $NO + Cl_2 \rightarrow Cl_2NO$ B) $2NO + Cl_2 \rightarrow 2ClNO$
C) $N_2 + O_2 + Cl_2 \rightarrow 2ClNO$ D) $NO + Cl \rightarrow ClNO$

Correct Answer: B
Level of Difficulty: Moderate/Challenging
Keyword(s): Visualizing reactions using balanced equations
Histogram:

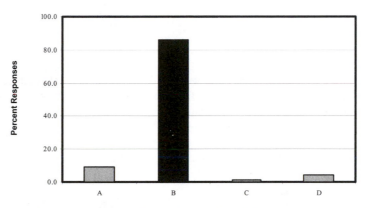

Comments: We were pleased that 86% of the students selected the correct answer to this question on the first hour exam. Reinforcement of atomic/molecular views of compounds and reactions helps with learning and retention.

9. What is the sum of the coefficients in the balanced equation using smallest integer coefficients for the reaction $B_2O_3(s) + NaOH(aq) \rightarrow Na_3BO_3(aq) + H_2O(l)$?

A) 9   B) 10   C) 12   D) 15

Correct Answer: C
Level of Difficulty: Moderate
Keyword(s): Balancing equations
Comments: As we expected, 92% of the students answered correctly as the formulas were provided. Most students appear to have learned this skill from high school chemistry.

10. What is the sum of the coefficients in the balanced equation using smallest integer coefficients when aluminum reacts with nitric acid (HNO_3) to form aluminum nitrate and hydrogen gas?

A) 8   B) 11   C) 13   D) 16

Correct Answer: C
Level of Difficulty: Moderate/Challenging
Keyword(s): Writing formulas and balancing equations

Histogram:

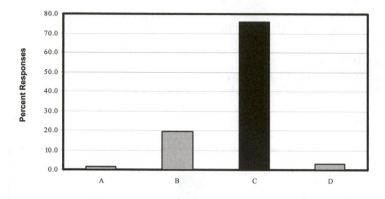

Comments: Given the formula of nitric acid, we hoped that students would deduce the negative charge on the nitrate ion and write the correct formula for aluminum nitrate. This question emphasizes the difference between aluminum metal and aluminum in the ionic compound. We share these ideas with our students after the histogram is projected.

11. Examine the molecular view of the reaction between AB and B_2 in the gas phase:

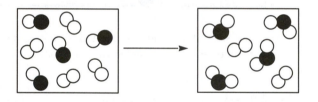

Select the correct statement about this reaction.

        A) The balanced equation for the reaction is $AB + B_2 \rightarrow AB_3$.
        B) AB and B_2 are present in stoichiometric amounts at the start of the reaction.
        C) AB is the limiting reactant.
        D) The product of the reaction is A_2B.

Correct Answer: C
Level of Difficulty: Challenging
Keyword(s): Balanced equations, stoichiometry, and limiting reactant

Note: Histogram and comments for this question appear on the next page.

Histogram:

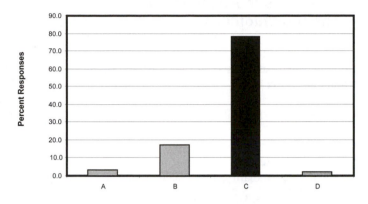

Comments: This question was presented on the first hour exam to reinforce the limiting reactant concept at the atomic/molecular level; 78% answered correctly. About 17% selected distractor (B); perhaps these students assumed the 1:1 ratio of reactants was the stoichiometric ratio and did not arrive at the correct balanced equation based on the formula of the product. We would definitely use this question in class in future lectures.

12. The formation of $NH_3(g)$ from $N_2(g)$ and $H_2(g)$ occurs in 85.0% yield. How many grams of ammonia would be experimentally obtained when 12.0 g of H_2 reacts with excess N_2?

A) 57.2 g   B) 66.9 g   C) 71.5 g   D) 83.8 g

Correct Answer: A
Level of Difficulty: Challenging
Keyword(s): Balanced equations and stoichiometry; theoretical, actual (or experimental), and percent yields
Histogram:

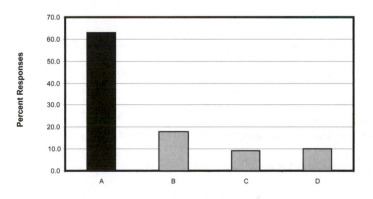

Comments: The use of a few quantitative questions in lecture is strongly encouraged although this question was given on the first hour exam. Students struggle with questions like this one; only 63% selected the correct answer.

CHAPTER 11 Reactions in Aqueous Solution, Molarity, and Stoichiometry

1. If 3.423 g of aluminum sulfate are present in 1.00×10^2 mL of the solution, what is the sulfate ion concentration? Formula unit mass for $Al_2(SO_4)_3$ is 342.3 g/mol.

A) 0.100 M   B) 0.200 M   C) 0.300 M   D) 0.500 M

Correct Answer: C
Level of Difficulty: Easy/Moderate
Keyword(s): Strong electrolytes, molarity, and concentrations of ions in aqueous solutions
Histogram:

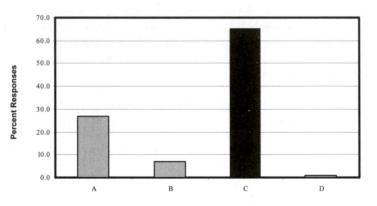

Comments: Calculating the concentrations of ions in solution poses a challenge to some students. It would be helpful to use cartoons to represent what happens to ionic compounds in solution, as well as the stoichiometry of the dissociation process. About 65% selected the correct answer to this question.

2. What is the total concentration of ions in 0.10 M iron(III) sulfate solution?

A) 0.10 M   B) 0.20 M   C) 0.30 M   D) 0.50 M

Correct Answer: D
Level of Difficulty: Easy/Moderate
Keyword(s): Strong electrolytes, molarity, and concentrations of ions in aqueous solutions

Note: Histogram and comments for this question appear on the next page.

Histogram:

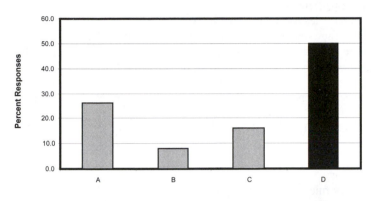

Comments: The distribution of responses confirms the lack of understanding of ionic concentrations in electrolyte solutions. Only 50% selected the correct answer.

3. Calculate the number of chloride ions in 52 L of 2.3 M copper(II) chloride solution.

A) 5.3×10^{22} B) 2.7×10^{25} C) 7.2×10^{25} D) 1.4×10^{26}

Correct Answer: D
Level of Difficulty: Moderate
Keyword(s): Molarity and concentrations of ions in solution
Histogram:

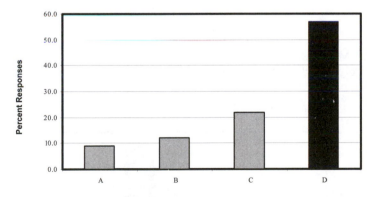

Comments: The correct answer was selected by 57% of the students while distractor (C) was chosen by 22% who may have neglected the formation of 2 moles of chloride ions for each mole of the compound. This question offers another opportunity to reinforce the bonding in the ionic compound to explain the ions formed.

4. Using solubility rules, select the compound that is least soluble in water.

A) $Pb(NO_3)_2$ B) $(NH_4)_2CO_3$ C) K_2SO_4 D) $Ca_3(PO_4)_2$

Correct Answer: D
Level of Difficulty: Easy/Moderate
Keyword(s): Solubility rules for ionic compounds
Comments: We encourage students to pay attention to the solutions used in the laboratory to become familiar with soluble salts. The solubility rules are usually provided on the cover page of the exam and the emphasis is on correct interpretation of the rules.

5. Predict the identity of the precipitate formed when solutions of Na_2CO_3 and $CaCl_2$ are mixed.

 A) Na_2CO_3 B) $CaCl_2$ C) $CaCO_3$ D) NaCl

Correct Answer: C
Level of Difficulty: Easy
Keyword(s): Solubility rules for ionic compounds
Comments: We were quite surprised when 22% of the students selected NaCl as the precipitate. None of the students selected distractors (A) or (B) confirming that they knew the difference between reactants and products.

6. An aqueous solution of H_2SO_4 is added to aqueous $Ba(OH)_2$. The reaction is monitored using a conductivity tester. Predict the correct statement(s).

 I) Both H_2SO_4 and $Ba(OH)_2$ are strong electrolytes.
 II) This is a neutralization reaction.
 III) This is a precipitation reaction.
 IV) The light bulb will glow at the neutralization point.

 A) II B) I and II C) I, II and III D) I, II, III and IV

Correct Answer: C
Level of Difficulty: Moderate/Challenging
Keyword(s): Electrolytes and conductivity, solubility rules, precipitation, and acids, bases, and neutralization
Histogram:

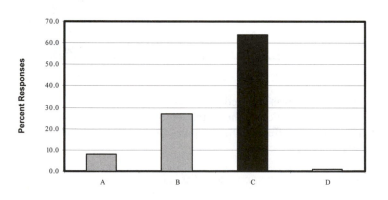

Comments: This is a very effective lecture demonstration to combine several concepts about electrolytes, conductivity, solubility and precipitation, and acid-base reactions. While 64% selected all correct statements, 27% did not recognize that precipitation would occur.

7. What happens to the number of moles of $C_{12}H_{22}O_{11}$ (sucrose) when a 0.20 M solution is diluted to a final concentration of 0.10 M?

> A) The number of moles of $C_{12}H_{22}O_{11}$ decreases.
> B) The number of moles of $C_{12}H_{22}O_{11}$ increases.
> C) The number of moles of $C_{12}H_{22}O_{11}$ does not change.
> D) There is insufficient information to answer the question.

Correct Answer: C
Level of Difficulty: Easy/Moderate
Keyword(s): Solute, solvent, solution, mole concept, and dilution
Histogram:

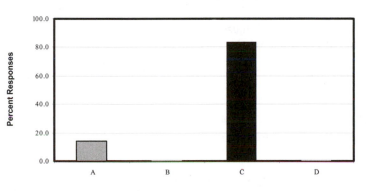

Comments: Understanding dilution at the particulate level is important when working with solutions. About 14% selected distractor (A) while 84% responded correctly. Some students confuse moles of solute with concentration.

8. Select the correct statement pertaining to solutions.

> A) The units of molarity are moles of solute per liter of solvent.
> B) The number of moles of solute does not change when a concentrated solution
>   is diluted.
> C) A 0.10 M NaCl solution contains 58.5 g of NaCl dissolved in one liter of water.
> D) A 0.10 M NaCl solution is a heterogeneous mixture.

Correct Answer: B
Level of Difficulty: Moderate
Keyword(s): Solutions, molarity, and dilution

Histogram:

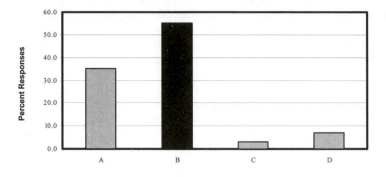

Comments: Students were encouraged to pay attention to detail such as the definition of molarity in terms of the volume of the solution, a necessary skill when preparing standard solutions; 35% selected distractor (A) while 55% answered correctly.

9. How many grams of $CuSO_4 \cdot 5H_2O$ must you weigh out in order to prepare 100.0 mL of a 0.10 M $CuSO_4$ solution? (MM: $CuSO_4 \cdot 5H_2O$ = 250 g/mol)

        A) 250 g   B) 25 g   C) 2.5 g   D) 0.25 g

Correct Answer: C
Level of Difficulty: Moderate
Keyword(s): Preparation of standard solutions
Histogram:

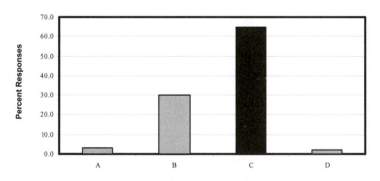

Comments: This question may be linked to a lecture demonstration to emphasize important details and laboratory skills.

10. How many mL of commercial HCl (12.0 M) should you use to prepare 240.0 mL of 0.10 M HCl?

        A) 1.2×10^{-4} L   B) 1.2×10^{-3} L
        C) 2.0×10^{-3} L   D) 2.4×10^{-3} L

Correct Answer: C
Level of Difficulty: Easy/Moderate
Keyword(s): Dilution
Comments: We use this question to reinforce the concept that the number of moles of solute remains unchanged on dilution. About 92% responded correctly. In our experience, students feel more comfortable memorizing the equation $M_1V_1 = M_2V_2$ without spending much time understanding the basis of this relationship. Consequently, they incorrectly use this equation for all acid-base titrations regardless of the stoichiometry.

11. What is the oxidation number of phosphorus in the phosphate ion?

   A) −5   B) −3   C) +3   D) +5

Correct Answer: D
Level of Difficulty: Easy
Keyword(s): Rules for oxidation numbers
Comments: This is a good opportunity to refer to the periodic table when explaining oxidation numbers. About 84% answered correctly.

12. Which of these is a redox reaction?

$$A) \ 2\ NaOH + H_2SO_4 \rightarrow Na_2SO_4 + 2\ H_2O$$
$$B) \ 2\ AgNO_3 + Cu \rightarrow Cu(NO_3)_2 + 2\ Ag$$
$$C) \ Pb(NO_3)_2 + 2\ KI \rightarrow PbI_2 + 2\ KNO_3$$
$$D) \ CaCO_3 + 2\ HCl \rightarrow CaCl_2 + CO_2 + H_2O$$

Correct Answer: B
Level of Difficulty: Easy
Keyword(s): Oxidation-reduction reactions, changes in oxidation numbers.
Comments: Students grasp this concept quite easily; 92% responded correctly.

CHAPTER 12 Thermodynamics and Thermochemistry

1. Which of these changes always results in an increase in the internal energy of the system?

        A) The system absorbs heat and does work on the surroundings.
        B) The system releases heat and does work on the surroundings.
        C) The system absorbs heat and has work done on it by the surroundings.
        D) The system releases heat and has work done on it by the surroundings.

Correct Answer: C
Level of Difficulty: Easy
Keyword(s): System and surroundings, internal energy, and heat and work
Comments: Students grasp this concept without too much difficulty; 99% answered correctly.

2. A system releases 300 J of heat and has 650 J of work done on it by the surroundings. What is the change in the internal energy of the system?

        A) +950 J   B) +350 J   C) −350 J   D) −950 J

Correct Answer: B
Level of Difficulty: Easy
Keyword(s): System and surroundings, heat, work, and internal energy, and first law of thermodynamics
Comments: Using the relationship $\Delta E = q + w$ and the sign convention, most students understand these concepts without much trouble. About 97% selected the correct answer.

3. Calculate the work done when a gas, at P = 1.0 atm, has a volume change from 22.0 L to V_{final} = 2.0 L, given 1 L·atm = 101.3 J.

        A) −2.0 kJ   B) −20. J   C) +20. J   D) +2.0 kJ

Correct Answer: D
Level of Difficulty: Moderate
Keyword(s): Relating work to volume changes at constant pressure

Note: Histogram and comments for this question appear on the next page.

Histogram:

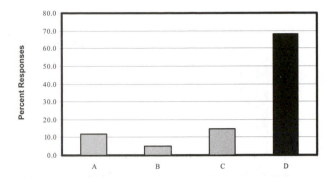

Comments: Only 68% responded correctly to this question. It appears that the major problem was keeping track of units.

4. Which property of a system is <u>not</u> a state function?

    A) Enthalpy, H   B) Heat, q   C) Internal energy, E   D) Pressure, P

Correct Answer: B
Level of Difficulty: Moderate/Challenging
Keyword(s): State functions, heat, work, pressure, enthalpy and internal energy
Histogram:

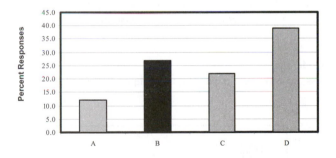

Comments: The histogram shows a distribution of responses; many students have a difficult time understanding the concept of state functions. Only 27% selected the correct answer.

5. Select the correct statement about combustion reactions.

        A) The products are relatively more stable than the reactants.
        B) These reactions are always endothermic.
        C) The reactants contain much stronger bonds compared to the products.
        D) Oxygen gas is a product of the reaction.

Correct Answer: A
Level of Difficulty: Moderate
Keyword(s): Combustion reactions, exothermic and endothermic processes, and energy profiles
Histogram:

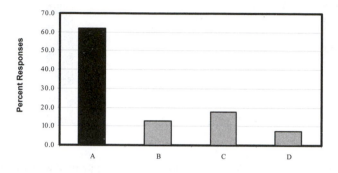

Comments: We encourage students to draw energy/enthalpy diagrams for exothermic and endothermic processes to understand relative stabilities of reactants and products. This skill is extremely useful in other chemistry courses. About 63% selected the correct answer.

6. The specific heat capacities of gold and copper are 0.129 J/g·°C and 0.387 J/g·°C, respectively. At 25°C, 230 joules of heat is added to 10.0-g samples of pure gold and copper. Select the correct statement about temperature changes. (Note: You can answer this question without detailed calculations.)

A) The temperature of the gold sample will rise higher than that of the copper sample.
B) The temperature of the copper sample will rise higher than that of the gold sample.
C) Both samples will experience the same increase in temperature.
D) Both samples will experience the same decrease in temperature.

Correct Answer: A
Level of Difficulty: Moderate
Keyword(s): Heat, mass, specific heat capacity, and temperature change
Histogram:

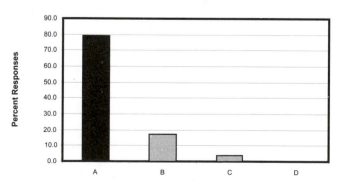

Comments: This question was presented on an hour exam and students were encouraged to answer the question without using detailed calculations. A similar question was presented in lecture. Many students resort to algorithmic problem solving approaches even when the answer can be deduced without calculations. The correct answer was selected by 79% of the students while 17% picked distractor (B).

7. Calculate the quantity of heat released when 4.00 g of $CH_4(g)$ are burned in excess oxygen according to the reaction: $CH_4(g) + 2O_2(g) \rightarrow CO_2(g) + 2H_2O(g)$; $\Delta H^0_{rxn} = -802.2$ kJ.

A) 802.2 kJ   B) 401.1 kJ   C) 200.6 kJ   D) 100.3 kJ

Correct Answer: C
Level of Difficulty: Moderate
Keyword(s): Thermochemical equations
Comments: In this question, the term "heat released" implies exothermicity; hence, signs are not shown with the numerical answers. About 94% responded correctly in lecture.

8. Given $\Delta H = -90.25$ kJ for the reaction: $NO(g) \rightarrow \frac{1}{2}N_2(g) + \frac{1}{2}O_2(g)$, calculate ΔH for the reaction: $N_2(g) + O_2(g) \rightarrow 2NO(g)$

A) -180.50 kJ   B) -90.25 kJ   C) $+90.25$ kJ   D) $+180.50$ kJ

Correct Answer: D
Level of Difficulty: Easy/Moderate
Keyword(s): Thermochemical equations
Comments: Students use thermochemical equations with some ease; 94% responded correctly to this question.

9. For which reaction is $\Delta H^0_{rxn} = \Delta H^0_f$?

A) $2H_2(g) + O_2(g) \rightarrow 2H_2O(l)$ B) $2C(graphite) + H_2(g) \rightarrow C_2H_2(g)$
C) $C(diamond) + O_2(g) \rightarrow CO_2(g)$ D) $NO(g) + \frac{1}{2}O_2(g) \rightarrow NO_2(g)$

Correct Answer: B
Level of Difficulty: Moderate
Keyword(s): Enthalpy of formation

Note: Histogram and comments for this question appear on the next page.

Histogram:

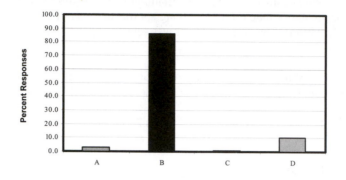

Comments: Questions dealing with ΔH^0_f can sometimes be challenging; however, 86% of the students responded correctly with peer collaboration.

10. Which statement is correct?

A) ΔH is a measure of heat absorbed or released by the system for a process carried out at constant volume.
B) According to sign convention, a positive q signifies that heat flows from the system to the surroundings.
C) The standard enthalpy of formation of C(*diamond*) at 25°C and 1 atm is zero.
D) The standard enthalpy of formation of $CH_4(g)$ is not the same as the standard enthalpy of combustion of $CH_4(g)$.

Correct Answer: D
Level of Difficulty: Moderate/Challenging
Keyword(s): Heat, enthalpy changes, and ΔH^0_f and ΔH^0_{comb}
Histogram:

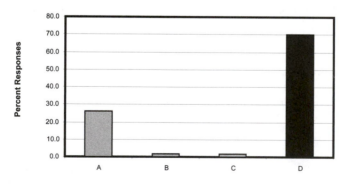

Comments: This question tests several concepts covered in the chapter on thermodynamics and thermochemistry and may be used to summarize concepts.

11. One mole of $N_2(g)$ is confined within a cylinder (fitted with a movable piston) at 25°C and heated to 275°C at a constant pressure of 0.975 atm. Calculate the work done (including sign) in joules. Assume that all of the energy is used to do work.

A) -2.08×10^3 J   B) -7.52×10^2 J   C) +71.3 J   D) +20.5 J

Correct Answer: A
Level of Difficulty: Challenging
Keyword(s): Gas expansion, heat and work
Histogram:

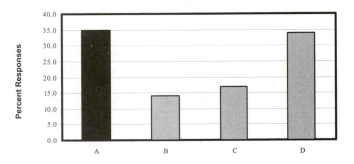

Comments: The histogram shows that the students were challenged by this question; 35% selected the correct answer while 34% picked distractor (D), an incorrect answer. This provided an opportunity to review the concept while working through the problem.

12. Calculate the standard enthalpy of formation of $CCl_4(g)$ using the thermochemical equations shown below:

$$CCl_4(g) + 4HCl(g) \rightarrow CH_4(g) + 4Cl_2(g) \qquad \Delta H^0 = +397.0 \text{ kJ}$$
$$\tfrac{1}{2}H_2(g) + \tfrac{1}{2}Cl_2(g) \rightarrow HCl(g) \qquad \Delta H^0 = -92.31 \text{ kJ}$$
$$C(graphite) + 2 H_2(g) \rightarrow CH_4(g) \qquad \Delta H^0 = -74.81 \text{ kJ}$$

A) -379.5 kJ/mol   B) -102.6 kJ/mol   C) $+167.1$ kJ/mol   D) $+227.9$ kJ/mol

Correct Answer: B
Level of Difficulty: Moderate/Challenging
Keyword(s): Hess's law, and standard enthalpy of formation

Note: Histogram and comments for this question appear on the next page.

Histogram:

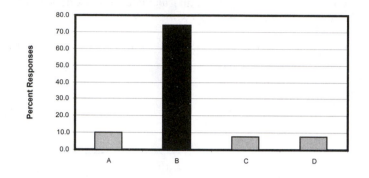

Comments: Most students can solve problems using Hess's law without any help. This question adds a challenge by integrating standard enthalpy of formation; 74% selected the correct answer.

13. Enthalpies of formation data are not always experimentally easy to obtain. However, enthalpies of combustion data are readily available. Calculate the enthalpy of formation of methane from the combustion data provided.

$$C(graphite) + O_2\ (g) \rightarrow CO_2\ (g) \qquad \Delta H^0_{comb} = -393.5\ kJ$$
$$H_2\ (g) + {}^1/_2O_2\ (g) \rightarrow H_2O(l) \qquad \Delta H^0_{comb} = -285.8\ kJ$$
$$CH_4\ (g) + 2O_2\ (g) \rightarrow CO_2\ (g) + 2H_2O(l) \qquad \Delta H^0_{comb} = -890.3\ kJ$$

A) −19.4 kJ/mol   B) −74.8 kJ/mol
C) −221.9 kJ/mol   D) −296.0 kJ/mol

Correct Answer: B
Level of Difficulty: Easy/Moderate
Keyword(s): Hess's law
Comments: This question was successfully completed by 83% of the students. The goal was to help students differentiate between enthalpies of formation and enthalpies of reactions such as combustion.

CHAPTER 13 Gases

1. Mercury (Hg) is the liquid of choice in a barometer. At sea level, 1 atm = 76 cmHg; the density of Hg is 13.5 g/mL How would the height of a water column at sea level compare with the height of a Hg column?

        A) Both columns would have the same height.
        B) The water column would be higher.
        C) The water column would be shorter.

Correct Answer: B
Level of Difficulty: Moderate
Keyword(s): Barometer and atmospheric pressure
Histogram:

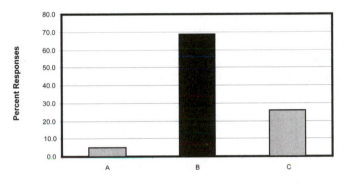

Comments: Students get a better appreciation for the high density of mercury and its practical application in barometers and manometers.

2. Observe the readings on the pressure gauge as the air in the syringe is compressed and then allowed to expand. What is the relationship between the volume (V) and pressure (P) of a gas?

        A) V is directly proportional to P.
        B) V is inversely proportional to P.
        C) There is no relationship between V and P.

Correct Answer: B
Level of Difficulty: Easy
Keyword(s): Pressure and volume relationship
Comments: This is an effective lecture demonstration. We usually remind students that the amount of the gas sample and temperature are held constant.

3. Imagine you are blowing up balloons in a room at some ambient temperature (and constant pressure). What do you think is the relationship between the volume of the balloon and the amount of gas (n = number of moles) in it?

> A) V is directly proportional to n.
> B) V is inversely proportional to n.
> C) There is no relationship between V and n.

Correct Answer: A
Level of Difficulty: Easy
Keyword(s): Relationship between volume and amount of gas
Comments: This can also be presented as a lecture demonstration; remind students that pressure and temperature do not change.

4. A 2.650-g sample of a gas occupies a volume of 428 mL at 0.9767 atm and 297.3 K. What is its molar mass?

> A) 97.5 g/mol   B) 123 g/mol   C) 155 g/mol   D) 186 g/mol

Correct Answer: C
Level of Difficulty: Moderate
Keyword(s): Ideal gas equation, and calculation of molar mass
Histogram:

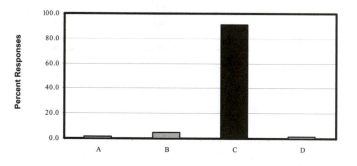

Comments: Students generally handle the ideal gas equation with confidence. Calculation of molar mass is an interesting application; 91% responded correctly.

5. Which sequence represents the gases in order of increasing density at STP?

> A) Fluorine < Carbon monoxide < Chlorine < Argon
> B) Carbon monoxide < Fluorine < Argon < Chlorine
> C) Argon < Carbon monoxide < Chlorine < Fluorine
> D) Fluorine < Chlorine < Carbon monoxide < Argon

Correct Answer: B
Level of Difficulty: Moderate/Challenging

Keyword(s): Applications of ideal gas equation, molar mass, density and STP
Histogram:

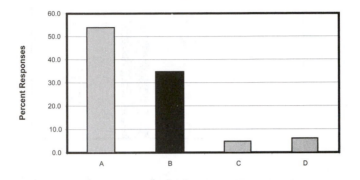

Comments: This question challenged some of the students; only 35% responded correctly. Students were required to rearrange the ideal gas equation and introduce density using mass and volume terms. This is a great example to review and reinforce several basic concepts.

6. What is the density of xenon gas at STP?

        A) 2.75 g/L   B) 4.10 g/L   C) 5.86 g/L   D) 7.22 g/L

Correct Answer: C
Level of Difficulty: Moderate
Keyword(s): Applications of ideal gas equation, molar mass, density and STP
Comments: Students performed much better on this question and 93% responded correctly.

7. A mixture consisting of 4.9 g CO and 8.5 g SO_2, two atmospheric pollutants, exerts a pressure of 0.761 atm when placed in a sealed container. What is the partial pressure of the SO_2 in this mixture?

        A) 0.13 atm   B) 0.18 atm   C) 0.33 atm   D) 0.43 atm

Correct Answer: C
Level of Difficulty: Moderate/Challenging
Keyword(s): Dalton's law of partial pressures and mole fractions

Note: Histogram and comments for this question appear on the next page.

Histogram:

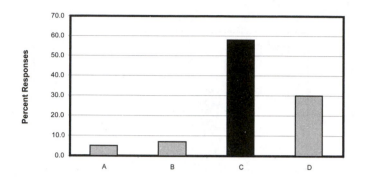

Comments: This question was presented on the second hour exam; 58% selected the correct answer. The question can be used in lecture by providing the molar masses of the gases.

8. A sample of an unknown gas effuses in 8.2 minutes. An equal volume of krypton in the same apparatus at the same temperature and pressure effuses in 4.0 minutes. What is the likely identity of the gas?

A) UF_6 B) Br_2 C) Ar   D) Ne

Correct Answer: A
Level of Difficulty: Moderate/Challenging
Keyword(s): Graham's law of diffusion/effusion
Histogram:

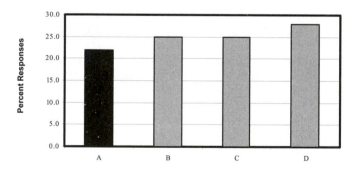

Comments: This question was presented on an exam and may take up time in lecture. However, it could be used in class by limiting time to 3 minutes and then using this example to discuss application of Graham's law. Students generally have a difficult time with this concept and the calculations.

9. Consider the figure below:

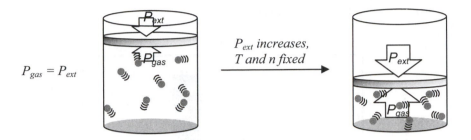

$P_{gas} = P_{ext}$

P_{ext} increases,
T and n fixed

The molecular description of ideal gas behavior depicted in this figure shows that

A) V is directly proportional to P.   B) V is inversely proportional to P.
C) V is directly proportional to T.   D) V is inversely proportional to T.

Correct Answer: B
Level of Difficulty: Easy/Moderate
Keyword(s): Kinetic molecular theory
Histogram:

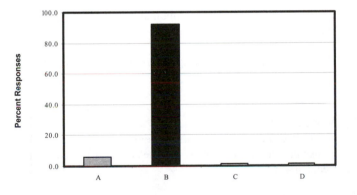

Comments: Molecular views play an important role in the discussion of kinetic molecular theory; 92% responded correctly to this question.

10. Three identical 5-liter flasks are filled with F_2, Cl_2, and Ar, respectively, at STP. Which statement is true?

A) The densities of all the gases are the same.
B) The velocities of the gas particles are the same in each of the three flasks.
C) The same number of gas particles is present in each flask.
D) The average kinetic energy of Cl_2 molecules is greater than that of either F_2 molecules or Ar atoms.

Correct Answer: C
Level of Difficulty: Challenging
Keyword(s): Kinetic molecular theory, velocities and kinetic energies
Histogram:

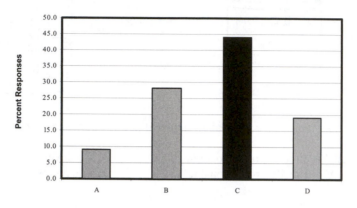

Comments: This is a challenging question and the histogram reflects a distribution of responses with 44% responding correctly. We have used this opportunity to discuss each statement and validate it as true or false.

11. A sample of an ideal gas is heated in a steel container from 25°C to 100°C. Which quantity will remain unchanged?

A) Average kinetic energy per gas particle   B) Collision frequency
C) Density   D) Pressure

Correct Answer: C
Level of Difficulty: Moderate/Challenging
Keyword(s): Kinetic molecular theory, average kinetic energy, and collision frequency
Comments: We would use this question to review all the concepts buried in the statements. About 81% selected the correct answer.

12. A hydrocarbon contains 85.7% carbon, by mass. If 14.0 g of the gaseous hydrocarbon occupies a volume of 6.12 L at 1 atm and 25°C, what is its molecular formula?

A) C_2H_4 B) C_4H_8 C) C_6H_{12} D) C_8H_{16}

Correct Answer: B
Level of Difficulty: Moderate/Challenging
Keyword(s): Ideal gas equation and molar mass, and empirical and molecular formulas

Histogram:

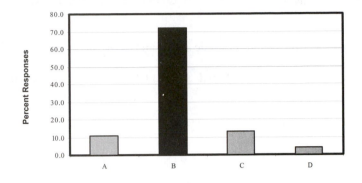

Comments: This question can be used to conclude the unit on gases and to make connections to concepts such as empirical and molecular formulas covered earlier in the course. About 72% responded correctly.

13. Select the correct statement about ideal gases.

    A) At constant T, the pressure is directly proportional to the volume for a given amount of gas.
    B) At constant P and T, the volume is inversely proportional to the number of moles of gas.
    C) At STP, one mole of any gas occupies a volume of 22.4 L.
    D) Deviations from ideal gas behavior are observed at low pressures and high temperatures.

Correct Answer: C
Level of Difficulty: Moderate/Challenging
Keyword(s): Gas laws, STP, and deviations from the ideal gas equation
Comments: This question may also be used to end the discussion on gases. Several concepts are tested and may be reviewed after the histogram has been projected.

CHAPTER 14 Atomic Structure and Quantum Mechanics

1. Which of these regions of the electromagnetic spectrum has the shortest wavelength?

   A) Gamma rays B) Infrared C) Ultraviolet D) Visible

Correct Answer: A
Level of Difficulty: Easy/Moderate
Keyword(s): Electromagnetic spectrum
Histogram:

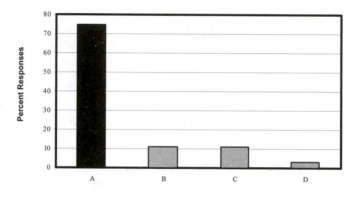

Comments: We require our students to memorize the regions of the electromagnetic spectrum with respect to relative energies. This question was presented on the hour exam and 75% showed mastery of the information; the percent of correct responses will probably be higher in lecture immediately after the information is presented.

2. What is the relationship between frequency and wavelength?

   A) Frequency is directly proportional to wavelength.
   B) Frequency is inversely proportional to wavelength.
   C) There is no relationship between frequency and wavelength.

Correct Answer: B
Level of Difficulty: Easy
Keyword(s): Frequency and wavelength
Comments: Our students often struggle with the concept of direct and inverse proportion and we try to reinforce this information whenever the opportunity arises. In this case, 97% responded correctly.

3. What is the wavelength of microwaves with a frequency of 2×10^{10} s^{-1}?

   A) 0.67 m B) 0.015 mm C) 663 nm D) 1.5 cm

Correct Answer: D
Level of Difficulty: Easy/Moderate
Keyword(s): Regions of electromagnetic spectrum and frequency and wavelength
Histogram:

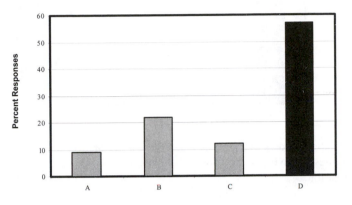

Comments: This should be a relatively simple calculation; yet, only 57% arrived at the correct answer. Perhaps, some students needed a review of the metric prefixes. It would be interesting to pose this question, then refrain from projecting the histogram, offer a review of the metric prefixes, allow the students to click in responses a second time, and compare the histograms.

4. Which salt emits light of the highest frequency in the flame test?

> A) Calcium chloride   B) Lithium chloride
> C) Potassium chloride   D) Sodium chloride

Correct Answer: C
Level of Difficulty: Easy/Moderate
Keyword(s): Emission of light, visible light, energy, and frequency
Comments: Colored flames produced by salts provide a popular lecture demonstration making connections to fireworks displayed on July 4th. We require our students to know the relative energies of the colored regions of the visible spectrum. This question actively engages students in the activity; 79% responded correctly.

5. One blue photon ($\lambda = 470$ nm) is sufficient to drive which reaction(s) below?

> A) Dissociation of $Cl_2(g)$ ($\Delta E^0 = 241$ kJ/mol)
> B) Dissociation of $H_2(g)$ ($\Delta E^0 = 435$ kJ/mol)
> C) Ionization of $Li(g)$ ($\Delta E^0 = 510$ kJ/mol)
> D) Both A) and B)

Correct Answer: A
Level of Difficulty: Moderate
Keyword(s): Energy of photon and bond dissociation, energy and wavelength

Histogram:

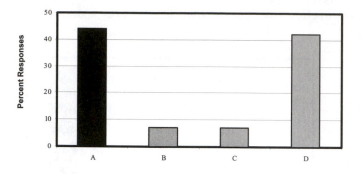

Comments: The histogram shows that 44% selected the correct answer and 42% selected distractor (D). This is a good question to make the connection between our discussion of energy and bond dissociation or ionization.

6. For which of the following species is the Bohr model applicable?

     A) $_2$He   B) $_4$Be$^+$ C) $_3$Li^{2+} D) $_1$H$^-$

Correct Answer: C
Level of Difficulty: Easy/Moderate
Keyword(s): Bohr model and one-electron systems
Comments: This question was intended to be a relatively simple one; only 56% responded correctly and 38% selected distractor (D). There appears to be some confusion about one-electron systems.

7. From which of these one-electron species is it most difficult to remove an electron?

     A) He$^+$ B) Li^{2+} C) Be^{3+} D) B^{4+}

Correct Answer: D
Level of Difficulty: Easy
Keyword(s): One-electron systems and energies of electrons
Comments: Students generally seemed confident responding to this question and 84% selected the correct response.

8. Which electronic transition in the hydrogen atom results in the emission of light of the longest wavelength?

     A) $n = 4$ to $n = 3$ B) $n = 1$ to $n = 2$
     C) $n = 1$ to $n = 6$ D) $n = 3$ to $n = 2$

Correct Answer: A
Level of Difficulty: Moderate
Keyword(s): Bohr model and electronic transitions in H atom

Histogram:

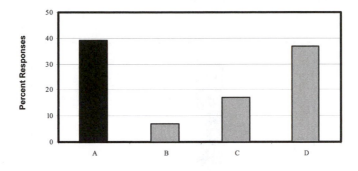

Comments: For several semesters, we have observed similar results both in class and on exams. After discussion of the process to arrive at the solution, we learned that many students do not draw a picture of the energy levels and overlook the important information that energy levels get closer as we move away from the nucleus. This is reflected in the histogram; 39% responded correctly and 37% selected distractor (D) corresponding to an emission. It also shows that 24% did not recognize the difference between absorption and emission of light.

9. The following set of quantum numbers is not allowed: $n = 3$, $l = 0$, $m_l = -2$. Assuming that the n and m_l values are correct, change the l value to an allowable combination.

   A) $l = 1$ B) $l = 2$ C) $l = 3$ D) $l = 4$

Correct Answer: B
Level of Difficulty: Easy
Keyword(s): Quantum numbers and allowed values
Comments: This question requires students to think about the limits of the quantum numbers; 96% responded correctly.

10. How many $4f$ orbitals are there?

   A) 4   B) 7   C) 8   D) 9

Correct Answer: B
Level of Difficulty: Easy
Keyword(s): Quantum numbers
Comments: Students are encouraged to use information about the allowed values of each quantum number; 80% responded correctly.

11. Which of these is a correct set of quantum numbers to describe a $5f$ orbital?

   A) $n = 5$, $l = 4$, $m_l = +3$ B) $n = 5$, $l = 3$, $m_l = +2$
   C) $n = 5$, $l = 2$, $m_l = +1$ D) $n = 5$, $l = 1$, $m_l = 0$

Correct Answer: B

Level of Difficulty: Easy
Keyword(s): Quantum numbers and allowed values
Histogram:

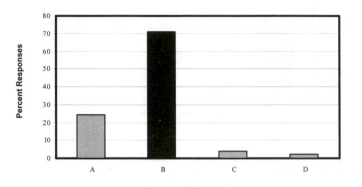

Comments: The correct answer was selected by only 71% of the students. About 24% selected distractor (A); these students appear to need more time to master the correct association of l values with labels such as s, p, d, f, g, etc.

12. In an atom, what is the maximum number of electrons that can have the quantum numbers $n = 4$ and $m_s = {}^1/_2$?

A) 4   B) 8   C) 16   D) 32

Correct Answer: C
Level of Difficulty: Easy/Moderate
Keyword(s): Quantum numbers and allowed values, and orbital occupancy
Histogram:

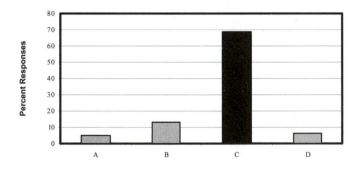

Comments: This question helps students to rationalize why there are three p orbitals or five d orbitals; 69% arrived at the correct answer.

13. Which of these is an incorrect orbital notation?

A) 3*p*   B) 2*p*   C) 5*s*   D) 2*d*

Correct Answer: D
Level of Difficulty: Easy
Keyword(s): Orbital notation
Comments: About 74% arrived at the correct answer while 23% selected distractor (C); perhaps the latter students need more time to gain confidence using labels such as *s*, *p*, *d*, *f*, *g*, etc. correctly.

14. Which statement about atomic orbitals and quantum numbers is correct?

A) The maximum number of orbitals with the quantum number $n = 3$ is 18.
B) There are five 2*d* orbitals.
C) The angular momentum quantum number is related to the shape of the orbital.
D) A 4*f* orbital is not possible.

Correct Answer: C
Level of Difficulty: Easy/Moderate
Keyword(s): Quantum numbers and allowed values, and orbital labels and shapes
Histogram:

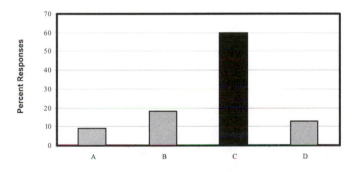

Comments: Only 60% answered correctly; 18% selected distractor (B) and 13% selected distractor (D) showing that students need more time to learn this information and to use it correctly.

15. What is the full set of valid quantum numbers for the electron lost when a potassium atom ionizes?

A) $n = 4, l = 0, m_l = 0, m_s = +^1/_2$ B) $n = 4, l = 1, m_l = +1, m_s = +^1/_2$
C) $n = 3, l = 1, m_l = +1, m_s = +^1/_2$ D) $n = 3, l = 0, m_l = +1, m_s = +^1/_2$

Correct Answer: A
Level of Difficulty: Moderate
Keyword(s): Quantum numbers for specific electrons in atoms

Histogram:

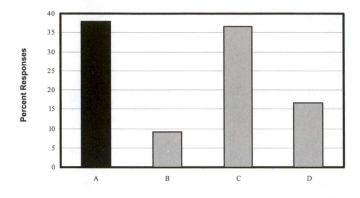

Comments: The histogram shows that 38% interpreted the question correctly and identified the quantum numbers for the 4s electron. About 37% selected distractor (C) with the quantum numbers for the 3p electron and appear to have misunderstood the question.

CHAPTER 15 Electron Configurations and Periodic Trends

1. What is the electron configuration of potassium (Z=19)?

A) $1s^22s^22p^63s^23p^63d^1$

B) $1s^22s^22p^63s^23p^64s^1$

C) $1s^22s^22p^63s^23p^33d^4$

D) $1s^12s^12p^33s^13p^43d^1$

Correct Answer: B
Level of Difficulty: Easy
Keyword(s): Electron configuration
Comments: The correct answer was selected by 98% of the students.

2. How many unpaired electrons are there in the iron (Z=26) atom?

A) 0　　　　　B) 2　　　　　C) 4　　　　　D) 6

Correct Answer: C
Level of Difficulty: Easy/Moderate
Keyword(s): Electron configurations, d-transition series and Hund's rule
Comments: The correct answer was selected by 86% of the students. We use this opportunity to reinforce Hund's rule.

3. Which element has the highest number of unpaired electrons?

A) Ca　　　　　B) Co　　　　　C) Cr　　　　　D) Cu

Correct Answer: C
Level of Difficulty: Easy/Moderate
Keyword(s): Electron configurations, d-transition series and Hund's rule
Histogram:

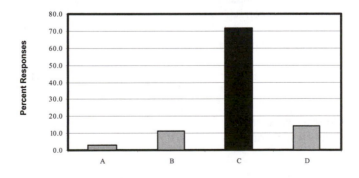

Comments: This question was presented on the hour exam to assess short-term retention and understanding of electron configurations and Hund's rule. The histogram shows that 28% did not respond correctly.

4. Select the electron configuration of $_{26}Fe^{3+}$, the ferric ion.

A) $[Ar]3d^64s^2$ B) $[Ar]3d^34s^2$ C) $[Ar]3d^6$ D) $[Ar]3d^5$

Correct Answer: D
Level of Difficulty: Easy/Moderate
Keyword(s): Electron configurations of ions
Comments: About 74% selected the correct answer; 23% selected distractor (B) that corresponds to the loss of three $3d$ electrons that is a common misconception.

5. Select the sequence of atoms that are correctly listed in order of increasing size.

A) F < Br < Ge < K   B) Na < Al < P < S
C) Ba < Ca < Mg < Be   D) Cl < Si < C < B

Correct Answer: A
Level of Difficulty: Easy
Keyword(s): Periodic trends in atomic sizes
Comments: Correct response was selected by 90% of the students; many students are able to rationalize atomic size trends.

6. List these atoms in order of increasing first ionization energy: $_3Li$, $_{11}Na$, $_6C$, $_8O$, $_9F$

A) Li < Na < C < O < F   B) Na < Li < C < O < F
C) F < O < C < Li < Na   D) Na < Li < C < F < O

Correct Answer: B
Level of Difficulty: Easy
Keyword(s): Periodic trends in first ionization energy
Comments: About 97% responded correctly with peer collaboration.

7. Which atom has the highest second ionization energy?

A) $_{19}K$ B) $_{20}Ca$ C) $_{21}Sc$ D) $_{56}Ba$

Correct Answer: A
Level of Difficulty: Moderate/Challenging
Keyword(s): Second ionization energy

Histogram:

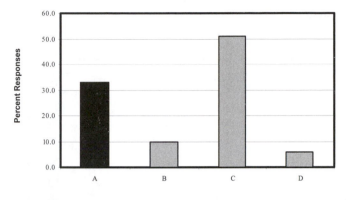

Comments: Only 51% selected the correct response. It is a good idea to review the definition of the second ionization energy and the process used to compare the relative energies of the relevant electrons that are lost during this ionization.

8. Select the smallest species.

A) Fe^{3+} B) Fe^{2+} C) Fe   D) K

Correct Answer: A
Level of Difficulty: Moderate
Keyword(s): Electron configurations of atoms and ions
Histogram:

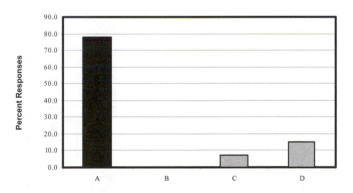

Comments: About 78% responded correctly and 15% selected distractor (D). It is a good idea to reinforce the concept that the outermost $4s$ electrons are lost before the inner $3d$ electrons.

9. Select the sequence that is listed in order of increasing size.

A) $_{18}Ar < _{19}K^+ < _{20}Ca^{2+} < _{17}Cl^- < _{16}S^{2-}$ B) $_{17}Cl^- < _{18}Ar < _{19}K^+ < _{20}Ca^{2+} < _{16}S^{2-}$
C) $_{19}K^+ < _{20}Ca^{2+} < _{17}Cl^- < _{16}S^{2-} < _{18}Ar$ D) $_{20}Ca^{2+} < _{19}K^+ < _{18}Ar < _{17}Cl^- < _{16}S^{2-}$

Correct Answer: D
Level of Difficulty: Moderate
Keyword(s): Sizes of atoms and ions
Histogram:

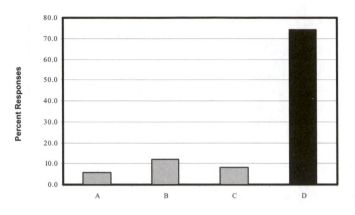

Comments: This is a great opportunity to reinforce the sizes of cations and anions compared to neutral atoms; 74% arrived at the correct answer.

10. Select the correct statement.

A) Li has a larger atomic radius than Cs.
B) Fe^{2+} has a smaller ionic radius than Fe^{3+}.
C) The ionization energy of H is greater than the first ionization energy of He.
D) The electronegativity of Mg is greater than that of Na.

Correct Answer: D
Level of Difficulty: Moderate
Keyword(s): Periodic trends in size, first ionization energy, and electronegativity
Histogram:

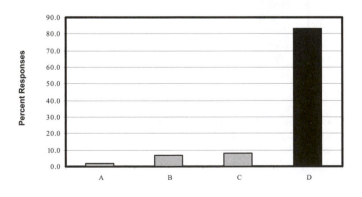

Comments: This question addresses trends in several properties and is a good way to summarize information; 83% responded correctly.

11. Which statement about periodic trends is correct?

    A) Metallic character decreases down a group.
    B) Anions are smaller than the corresponding neutral atoms.
    C) In most cases, energy is released when the first electron is added to an isolated gaseous atom.
    D) Most metals tend to form acidic oxides.

Correct Answer: C
Level of Difficulty: Moderate/Challenging
Keyword(s): Periodic trends in properties
Histogram:

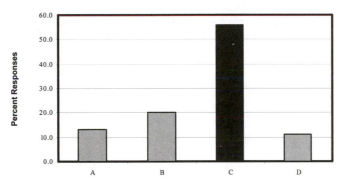

Comments: This question also addresses trends in several properties and is a good way to summarize information; only 56% responded correctly. The histogram shows a distribution of responses.

12. Which statement is correct?

    A) Atomic radius generally increases from left to right across a row (period).
    B) First ionization energy increases down a column (group).
    C) Second ionization energy is always smaller than the first ionization energy.
    D) The addition of an electron to a gaseous neutral atom is exothermic in most cases.

Correct Answer: D
Level of Difficulty: Moderate/Challenging
Keyword(s): Periodic trends in properties

Histogram:

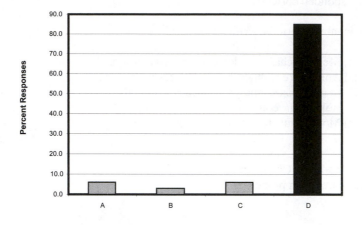

Comments: This question also addresses trends in several properties and is a good way to summarize information; 85% responded correctly.

CHAPTER 16 Chemical Bonding

1. Using Lewis structures, predict the formula for the binary ionic compound containing magnesium and nitrogen.

A) Mg_2N B) MgN_2 C) Mg_3N_2 D) Mg_2N_3

Correct Answer: C
Level of Difficulty: Easy
Keyword(s): Ionic bonding, Lewis structures
Comments: Lewis structures provide a rationale for the ratios in which metals and nonmetals combine to form ionic compounds; 85% responded correctly.

2. For each pair, choose the compound with the higher lattice energy:

I) LiF or KBr   II) NaCl or MgS

A) LiF and NaCl   B) LiF and MgS   C) KBr and NaCl   D) KBr and MgS

Correct Answer: B
Level of Difficulty: Moderate/Challenging
Keyword(s): Lattice energy, ionic charges, and ionic sizes
Histogram:

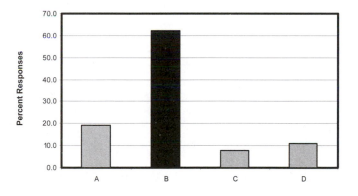

Comments: This is a moderately challenging concept for many students; 62% arrived at the correct answer. We generally spend time explaining the thought process in detail as we review lattice energy.

3. Arrange the compounds KI, $CaBr_2$, and CaS in order of increasing melting point using lattice energy as a guideline.

A) $CaS < CaBr_2 < KI$ B) $CaBr_2 < CaS < KI$

C) $KI < CaBr_2 < CaS$ D) $CaBr_2 < KI < CaS$

Correct Answer: C
Level of Difficulty: Moderate/Challenging
Keyword(s): Lattice energy, ionic charges, ionic sizes, and melting points
Histogram:

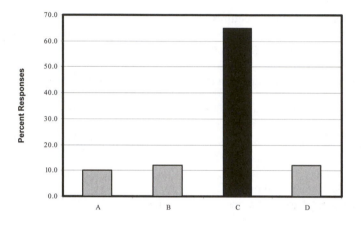

Comments: This histogram also confirms that lattice energy is a difficult topic for students; only 65% answered correctly.

4. How many double bonds are there in the hydrazine (N_2H_4) molecule?

A) 0   B) 1   C) 2   D) 3

Correct Answer: A
Level of Difficulty: Moderate
Keyword(s): Valence electrons, octet rule, and Lewis structure

Note: Histogram and comments for this question appear on the next page.

Histogram:

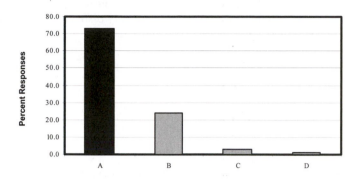

Comments: This should be an easy question if students keep track of the total number of valence electrons and apply the octet rule. However, we have found on several occasions that students tend to connect the nitrogen atoms by a double bond even in the molecular modeling exercise in the laboratory. This observation prompted this question. While 73% answered correctly, 24% selected distractor (B) for one double bond.

5. Identify the weakest covalent bond.

A) C–O   B) C–C   C) C=C   D) C≡C

Correct Answer: B
Level of Difficulty: Moderate
Keyword(s): Atomic sizes, bond lengths, electronegativity, and bond strengths
Comments: This question allows us to reinforce atomic sizes and electronegativity to predict bond lengths and strengths. This is an essential skill, especially in organic chemistry. Students recognized that multiple bonds are stronger than single bonds; 57% answered correctly while 35% selected distractor (A).

6. Identify the most polar covalent bond.

A) N–H   B) O–H   C) H–F   D) C–F

Correct Answer: C
Level of Difficulty: Easy/Moderate
Keyword(s): Electronegativity differences and bond polarities
Comments: About 93% selected the correct response to this question.

7. The best Lewis structure of HCN (carbon is the central atom) has

A) 1 nonbonding pair on carbon   B) 1 nonbonding pair on nitrogen
C) 2 nonbonding pairs on nitrogen   D) 2 nonbonding pairs on carbon

Correct Answer: B

Level of Difficulty: Moderate
Keyword(s): Valence electrons, multiple bonds, nonbonding electrons, and Lewis structures
Comments: Our experience shows that students have a relatively difficult time drawing correct Lewis structures and this weakness spills over into organic chemistry. About 76% arrived at the right answer.

8. Which of these is an exception to the octet rule?

  A) NF_3 B) H_3O^+ C) PCl_3 · D) IF_4^+

Correct Answer: D
Level of Difficulty: Moderate/Challenging
Keyword(s): Valence electrons, Lewis structures, octet rule and exceptions
Histogram:

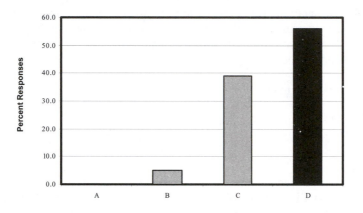

Comments: About 39% selected distractor (C) and 57% answered correctly even with peer collaboration.

9. For which molecule is the molecular geometry (shape) different from its electronic geometry?

  A) CH_4 B) NH_3 C) BF_3 D) PCl_5

Correct Answer: B
Level of Difficulty: Easy
Keyword(s): VSEPR theory, electronic and molecular geometry
Comments: This question appears to be quite straightforward; 93% selected the correct answer.

10. Predict the molecular shape (geometry) and bond angle of the sulfur dichloride, SCl_2, molecule.

  A) Linear, 180° B) Trigonal planar, 120°
  C) Trigonal pyramidal, < 109.5° D) Bent (V-shaped), <109.5°

Correct Answer: D
Level of Difficulty: Easy/Moderate
Keyword(s): VSEPR theory, molecular geometry, shape and bond angle
Histogram:

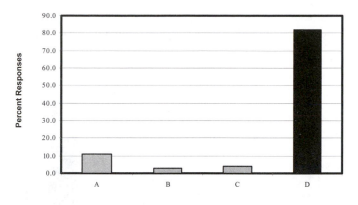

Comments: The histogram shows that 11% selected distractor (A) and 83% answered correctly. It appears that the former group did not consider the larger volume requirements of lone pairs and the consequent effect on bond angles.

11. Using VSEPR theory, predict the shape of IF_4^+.

A) Tetrahedral   B) Square planar   C) See-saw   D) Octahedral

Correct Answer: C
Level of Difficulty: Moderate
Keyword(s): VSEPR theory and shapes
Comments: The tetrahedral and square planar distractors were selected by 16% and 19%, respectively; 60% answered correctly. Perhaps, mastery of this concept takes more time.

12. Which compound is represented by this molecular view of the gas phase?

A) NO_2 B) SCl_2 C) CO_2 D) SO_2

Correct Answer: C
Level of Difficulty: Easy/Moderate
Keyword(s): VSEPR theory, shapes, and molecular view
Comments: This question was presented on the hour exam to test retention; while 85% answered correctly, 10% selected distractor (B) compared to 11% who selected this incorrect response on the conceptual question (#10 above) about SCl_2 presented in lecture.

13. Select the polar molecule.

        A) BF_3 B) XeF_2 C) CS_2 D) BrF_3

Correct Answer: D
Level of Difficulty: Challenging
Keyword(s): VSEPR theory, shape, and polarity of bonds and molecules
Histogram:

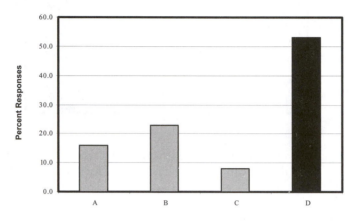

Comments: This question combines several concepts and our experience shows that students are challenged when asked to assign net polarity; only 53% answered correctly on the hour exam. As the histogram indicates, 16% selected distractor (A) and 23% selected distractor (B). This question can be modified for use in lecture by using only 1 example and including statements about shape, bond polarities, bond angles, and net polarity.

14. Given the data below, calculate the enthalpy of reaction when 1 mol Br_2 adds to C_2H_4 to form $C_2H_4Br_2$ (all reactants and products are gases).

Bond Type	C=C	C–C	C–H	C–Br	Br–Br
Bond Energy (kJ/mol)	614	347	413	276	193

        A) −92 kJ   B) −505 kJ   C) +184 kJ   D) +255 kJ

Correct Answer: A
Level of Difficulty: Challenging
Keyword(s): Lewis structures, bond dissociation energies, and enthalpy of reaction

Note: Histogram and comments for this question appear on the next page.

Histogram:

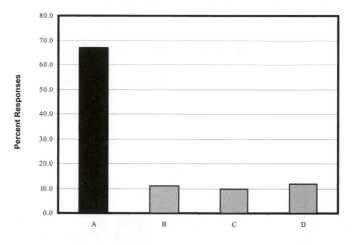

Comments: This question may be time-intensive for use in class; however, it is an excellent way to conclude the unit. Perhaps, it could be posted online prior to lecture (using the Just-In-Time Teaching philosophy) and then presented in class to assess understanding. We wrote this question for an hour exam and 67% answered correctly.

CHAPTER 17 Theories of Chemical Bonding

1. What is the hybridization of each carbon atom in acetic acid, CH_3COOH?

      A) sp, sp^2 B) sp^2, sp^2 C) sp^3, sp^2 D) sp^3, sp^3

Correct Answer: C
Level of Difficulty: Moderate/Challenging
Keyword(s): Valence bond theory and hybridization
Histogram:

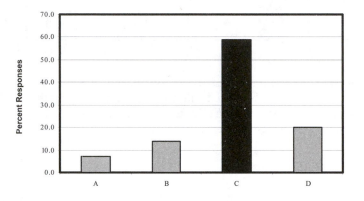

Comments: Our experience confirms that hybridization is a challenging concept for many students and we observe this in the organic chemistry courses as well. While 59% answered correctly in lecture, 20% thought that both carbon atoms were sp^3 hybridized. We find that students generally do better on these questions on hour exams after some drill and practice.

2. Select all the correct statements about CO_2.

      I) It is a linear molecule containing polar covalent bonds.
      II) It is a nonpolar molecule.
      III) The carbon atom in CO_2 is sp^2 hybridized.

      A) I and II   B) II and III   C) I and III   D) I, II, and III

Correct Answer: A
Level of Difficulty: Moderate
Keyword(s): Polarity of molecules, and hybridization

Note: Histogram and comments for this question appear on the next page.

Histogram:

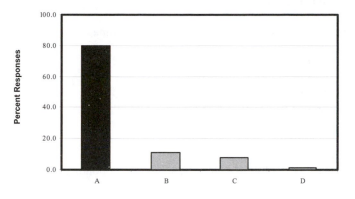

Comments: This question reviews some concepts covered earlier in the course and reinforces the idea that CO_2 is a nonpolar molecule due to its linear shape even though there are polar covalent bonds. This is also a good example of a molecule containing two cumulative double bonds that are best explained by a sp-hybridized central atom.

3. Select all the correct statements about chlorine trifluoride.

> I) The molecular geometry (shape) is trigonal bipyramidal.
> II) It is a polar molecule.
> III) The chlorine atom is sp^3d hybridized.

   A) I and II   B) II and III   C) I and III   D) I, II, and III

Correct Answer: B
Level of Difficulty: Challenging
Keyword(s): VSEPR theory and shape, polarity of molecule, and hybridization
Histogram:

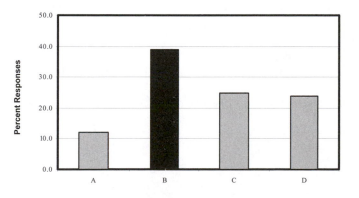

Comments: The histogram reflects the challenges that students faced with this question; only 39% answered correctly. We use this opportunity to engage students by reviewing electronic and

molecular geometries and the occupancy of the equatorial positions by the lone pairs. This leads to a T-shaped molecule and net polarity.

4. Identify the best set of bonds for the cyanate NCO^- and fulminate CNO^- ions.

> A) $[N\equiv C-O]^-$ & $[C=N=O]^-$
> B) $[N=C=O]^-$ & $[C=N=O]^-$
> C) $[N\equiv C-O]^-$ & $[C\equiv N-O]^-$
> D) $[N\equiv C-O]^-$ & $[C-N\equiv O]^-$

Correct Answer: C
Level of Difficulty: Moderate/Challenging
Keyword(s): Lewis structures and formal charges, and multiple bonding
Histogram:

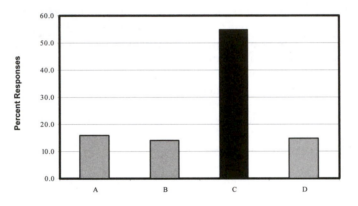

Comments: This histogram also shows a distribution of responses with 55% responding correctly.

5. Which designation best describes the hybrid orbitals on the central atom in SF_4?

> A) sp^2 B) sp^3 C) sp^3d D) sp^3d^2

Correct Answer: C
Level of Difficulty: Easy/Moderate
Keyword(s): Lewis structures and octet expansion, VSEPR theory, and hybridization

Note: Histogram and comments for this question appear on the next page.

Histogram:

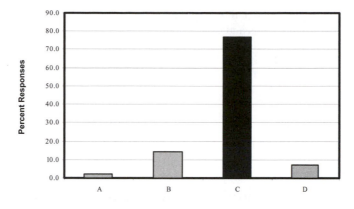

Comments: The students handled this question reasonably well as 77% answered correctly.

6. The hybridization of Xe in XeF$_4$ is

     A) sp^2.   B) sp^3.   C) sp^3d.   D) sp^3d^2.

Correct Answer: D
Level of Difficulty: Moderate
Keyword(s): Lewis structures and octet expansion, VSEPR theory, and hybridization
Histogram:

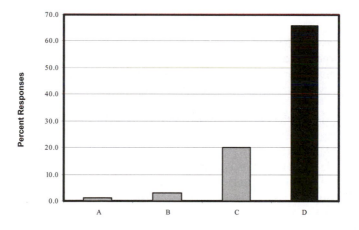

Comments: Students appear to have recognized octet expansion and the use of *d* orbitals in hybridization. Only 66% correctly identified the octahedral electronic geometry.

7. In CO_2, chemical bonding is best described in terms of

> A) one sigma bond and three pi bonds.
> B) two sigma bonds and two pi bonds.
> C) three sigma bonds and one pi bond.
> D) four sigma bonds.

Correct Answer: B
Level of Difficulty: Easy
Keyword(s): Sigma and pi bonds
Comments: About 94% selected the correct answer.

8. Which molecule contains one pi bond?

> A) C_2H_6 B) CH_3CH_2OH C) CH_3OCH_3 D) $(CH_3)_2CO$

Correct Answer: D
Level of Difficulty: Easy/Moderate
Keyword(s): Lewis structures, multiple bonding, and sigma and pi bonds
Comments: Condensed structural formulas were used in preparation for the unit on organic chemistry. About 86% selected the correct answer.

9. Which molecules contain one or more pi (π) bonds?

> A) C_2H_2 and C_2H_4 B) C_2H_2 and C_2H_6
> C) C_2H_4 and C_2H_6 D) C_2H_2, C_2H_4 and C_2H_6

Correct Answer: A
Level of Difficulty: Easy/Moderate
Keyword(s): Lewis structures and multiple bonding, sigma and pi bonds
Comments: These simple organic molecules are often used in lecture on several occasions. 82% selected the correct answer and 13% selected distractor (C).

10. Using MO theory, predict the potential existence of He_2 and/or He_2^+.

> A) He_2 and He_2^+ may exist.
> B) He_2 and He_2^+ may not exist.
> C) He_2 may exist and He_2^+ may not exist.
> D) He_2 may not exist and He_2^+ may exist.

Correct Answer: D
Level of Difficulty: Moderate
Keyword(s): Molecular orbital theory
Comments: Many general chemistry instructors do not cover MO theory while organic chemistry instructors use MO theory to discuss reactions, including acid-base reactions. This question was presented as a conceptual question in an organic chemistry class after a very brief introduction to MO theory and about 50% answered correctly.

11. Use molecular orbital theory to select the correct statement.

> A) C_2^- has a longer bond length than C_2^+.
> B) O_2^- is an unstable species.
> C) NO has a bond order of 2.5.
> D) O_2 is a diamagnetic molecule.

Correct Answer: C
Level of Difficulty: Challenging
Keyword(s): Molecular orbital theory, bond order, and magnetic properties
Histogram:

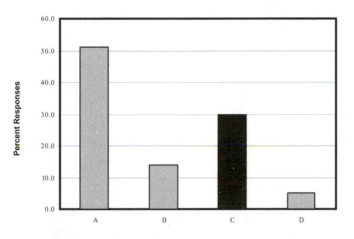

Comments: This question was presented on an hour exam and 51% answered correctly. The coverage of MO theory in first-semester general chemistry is a controversial issue and those of us who teach organic chemistry would prefer some basic introduction to MO theory in general chemistry.

CHAPTER 18 Organic Chemistry

1. Predict the molecular formula for the C_4 alkane based on the discussion of ethane and propane.

A) C_4H_4 B) C_4H_6 C) C_4H_8 D) C_4H_{10}

Correct Answer: D
Level of Difficulty: Easy
Keyword(s): Alkane formulas
Comments: Students grasped this concept quickly and 96% responded correctly. A variation of this question would be to ask students to identify the correct general formula for alkanes (C_nH_{2n+2}).

2. Select the molecular formula for the alkane $(CH_3)_2CH(CH_2)_3CH(CH_3)_2$.

A) C_7H_{14} B) C_7H_{16} C) C_9H_{18} D) C_9H_{20}

Correct Answer: D
Level of Difficulty: Moderate
Keyword(s): Condensed structural formulas
Histogram:

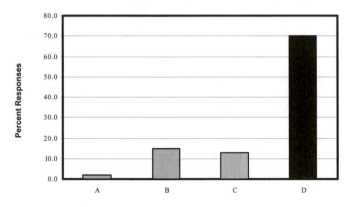

Comments: Our experience shows that students need drill and practice with condensed structural formulas to gain confidence in using these. 70% answered correctly on this example.

3. The name 3,3-dimethylbutane is incorrect. Select the correct IUPAC name of this compound.

A) hexane   B) 3,3-dimethylpentane
C) 2,2-dimethylbutane   D) 2,2-dimethylpropane

Correct Answer: C
Level of Difficulty: Easy/Moderate
Keyword(s): IUPAC nomenclature rules for alkanes

Comments: It is good practice to teach students to draw a structure based on the incorrect name and then apply the IUPAC rules to write the correct name. We were pleased when 87% responded correctly.

4. What is the maximum number of structural (or constitutional) isomers of C_5H_{12}?

A) 2   B) 3   C) 4   D) 5

Correct Answer: B
Level of Difficulty: Moderate
Keyword(s): Isomers, drawing structures of structural (or constitutional) isomers
Histogram:

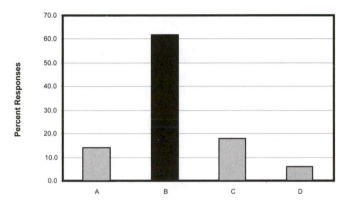

Comments: In our experience, both in general and organic chemistry, the concept of structural isomers challenges many students. Even with peer collaboration, only 62% arrived at the correct answer. When explaining the solution to this question, we usually write out the name for each structural isomer; if we have different representations of the same molecule, they have the same name and therefore represent a single compound.

5. Which of these is not a structural isomer of 2-methylhexane?

A) 2,4-dimethylpentane   B) 2,2-dimethylpentane
C) 2,2,3-trimethylbutane   D) 2,2,3,3-tetramethylbutane

Correct Answer: D
Level of Difficulty: Easy/Moderate
Keyword(s): Recognizing structural (or constitutional) isomers

Note: Histogram and comments for this question appear on the next page.

Histogram:

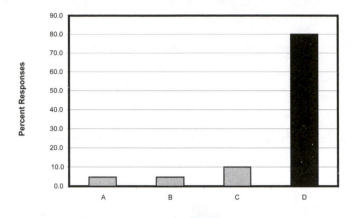

Comments: About 80% correctly recognized structural isomers when names were provided.

6. Which compound is not a structural (or constitutional) isomer of the other three?

A) 2-methylpropane   B) 2-methylbutane
C) 2,2-dimethylpropane   D) pentane

Correct Answer: A
Level of Difficulty: Easy/Moderate
Keyword(s): Recognizing structural (or constitutional) isomers
Histogram:

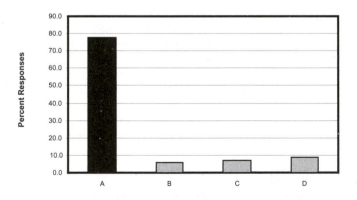

Comments: About 78% responded correctly confirming that the recognition of structural (or constitutional) isomers is not the challenge; drawing all possible structural isomers given the formula is the greater challenge.

7. How many different isomers (structural and geometric) are possible with the molecular formula C_4H_8?

A) 3 B) 4 C) 5 D) 6

Correct Answer: D
Level of Difficulty: Challenging
Keyword(s): Structural and geometric isomers
Histogram:

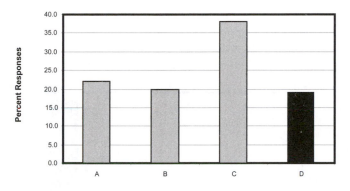

Comments: The histogram confirms the challenges of this question. Only 19% answered correctly. This is a good example to discuss in class and show effective approaches to answering questions like this without spending too much time. Cycloalkanes have the same general molecular formula C_nH_{2n} as alkenes.

8. Which of these molecules exhibit geometric (*cis-trans*) isomerism?

I) 2-methylpropene II) 2-butene III) 2-pentene

A) I and II B) I and III C) II and III D) I, II and III

Correct Answer: C
Level of Difficulty: Easy/Moderate
Keyword(s): Geometric or *cis-trans* isomerism
Comments: We were pleased when 91% responded correctly to this question that requires drawing correct structures and recognizing the substituents at the doubly-bonded carbon atoms.

9. Addition of Br_2 to an unknown compound produced 1,2-dibromobutane. The unknown compound was:

A) butane B) 1-butene C) 2-butene

Correct Answer: B
Level of Difficulty: Easy/Moderate
Keyword(s): Addition reactions of alkenes

Comments: This question appears to have been easy as 92% responded correctly.

10. Which of these molecules contain chiral carbon atoms?

I) $H_2N-CH(CH_3)-COOH$ (alanine)
II) $CH_3CH(OH)-COOH$ (lactic acid)
III) $HOH_2C-CH(OH)-COOH$ (glyceric acid)

A) I and II   B) I and III   C) II and III   D) I, II and III

Correct Answer: D
Level of Difficulty: Moderate/Challenging
Keyword(s): Chirality and chiral carbon atoms
Histogram:

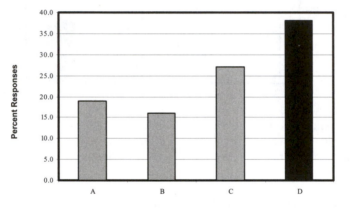

Comments: The histogram shows that our students had a difficult time with the identification of chiral carbon atoms in the condensed structural formulas shortly after the information was presented in lecture. Only 38% answered correctly.

11. Which molecule contains a chiral carbon atom?

A) $CH_3CH_2CH(Cl)CH_2CH_3$ B) $BrCH_2CH(Cl)CH_2Br$
C) $CH_3C(Br)(Cl)CH_3$ D) $CH_3CH(OH)COOH$

Correct Answer: D
Level of Difficulty: Moderate
Keyword(s): Chirality and chiral carbon atoms
Comments: This question was presented on the hour exam to test for understanding and retention after some drill and practice on assigned homework. 81% selected the right answer.

12. Identify the functional groups in this molecule.

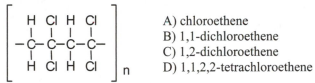

A) Alcohol, amide, carboxylic acid
B) Aldehyde, amine, ester
C) Alcohol, amine, carboxylic acid
D) Aldehyde, amide, ketone

Correct Answer: C
Level of Difficulty: Easy/Moderate
Keyword(s): Functional groups
Comments: About 94% answered correctly just after functional groups were discussed in class and the information was probably copied into students' notebooks.

13. The polymer poly(vinylidene chloride) is used for food wrap (e.g., Saran). Which monomer is used to make this polymer?

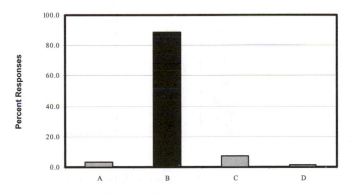

A) chloroethene
B) 1,1-dichloroethene
C) 1,2-dichloroethene
D) 1,1,2,2-tetrachloroethene

Correct Answer: B
Level of Difficulty: Moderate/Challenging
Keyword(s): Monomers and polymers
Histogram:

Comments: The representation of the repeating unit using the expanded structural formula may explain why 89% responded correctly compared to only 19% in a previous semester when the condensed structural formula $-(CH_2-CCl_2-CH_2-CCl_2)_n-$ was presented.

CHAPTER 19 Molecular Interactions and Properties
of Solids and Liquids

1. What physical change is represented by the equation $H_2O(g) \rightarrow H_2O(l)$?

A) Freezing   B) Condensation   C) Sublimation   D) Vaporization

Correct Answer: B
Level of Difficulty: Easy
Keyword(s): Phase changes
Comments: A quick review and reinforcement of phase transitions is highly recommended. 96% responded correctly to this question. A variation of this question would be a molecular view of the phase change.

2. Which phase change is an exothermic process?

A) Melting   B) Evaporation   C) Sublimation   D) Condensation

Correct Answer: D
Level of Difficulty: Easy/Moderate
Keyword(s): Enthalpy changes accompanying phase transitions
Histogram:

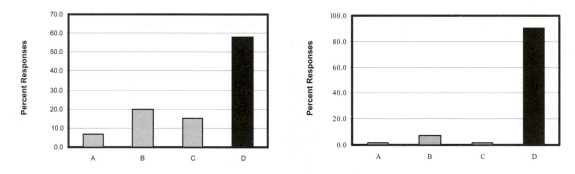

Comments: The histogram on the left represents responses by individual students without peer collaboration; only 58% responded correctly on their own. As shown in the histogram on the right, peer collaboration led to 90% correct responses. The first histogram was not displayed prior to peer collaboration.

3. Consider the process occurring in pure water at its boiling point. Which equation best represents this process?

A) $H_2O(l) \rightarrow H_2O(g)$ B) $2\ H_2O(l) \rightarrow 2\ H_2(g) + O_2(g)$
C) $H_2O(g) \rightarrow H_2O(l)$ D) $2\ H_2(g) + O_2(g) \rightarrow 2\ H_2O(l)$

Correct Answer: A
Level of Difficulty: Moderate
Keyword(s): Phase change in boiling water
Histogram:

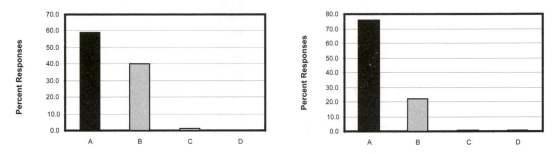

Comments: As in the previous question, peer collaboration (histogram on the right) increased the percent of correct responses from 59% to 76%. Our students prefer the peer collaboration approach and feel less pressured; however, they are always very cooperative when we request individual participation for data gathering. Both histograms confirm the misconception that hydrogen and oxygen gases are produced (distractor [B]) when water boils.

4. Consider the electrolysis of water. Which equation best represents the change that occurs?

A) $H_2O(l) \rightarrow H_2O(g)$ B) $2\ H_2O(l) \rightarrow 2\ H_2(g) + O_2(g)$
C) $H_2O(g) \rightarrow H_2O(l)$ D) $2\ H_2(g) + O_2(g) \rightarrow 2\ H_2O(l)$

Correct Answer: B
Level of Difficulty: Easy
Keyword(s): Electrolysis of water, chemical change
Comments: An overwhelming 98% answered correctly to this question. The lecture demonstration was performed after the histogram was displayed.

5. For the following pairs of substances, indicate which is more polarizable:

I) Br or Br⁻   II) propene or propane

A) Br and propene   B) Br and propane
C) Br⁻ and propene   D) Br⁻ and propane

Correct Answer: C
Level of Difficulty: Moderate
Keyword(s): Polarizability and dispersion forces

Note: Histogram and comments for this question appear on the next page.

Histogram:

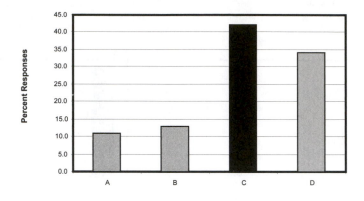

Comments: Our experience shows that students have a difficult time grasping the concept of polarizability, and the histogram displays a distribution of responses supporting our observation.

6. Which pure substances will not form hydrogen bonds?

I) CH_3CH_2OH II) CH_3OCH_3 III) $H_3C-NH-CH_3$ IV) CH_3F

A) I and II   B) I and III   C) II and III   D) II and IV

Correct Answer: D
Level of Difficulty: Moderate
Keyword(s): Hydrogen bonding
Histogram:

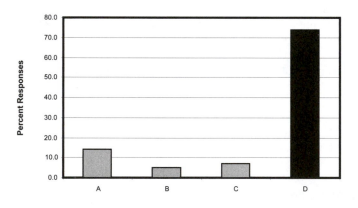

Comments: Many students have been exposed to the concept of hydrogen bonding especially in biology classes. 74% of the students responded correctly to this question.

7. Select the answer where the compounds are listed in order of increasing melting point.

A) $Na_2O < NaI < C_2H_5OH < C_3H_8$ B) $C_3H_8 < C_2H_5OH < NaI < Na_2O$
C) $NaI < Na_2O < C_3H_8 < C_2H_5OH$ D) $C_2H_5OH < C_3H_8 < Na_2O < NaI$

Correct Answer: B
Level of Difficulty: Moderate/Challenging
Keyword(s): Intermolecular forces and melting points of ionic and covalent compounds
Histogram:

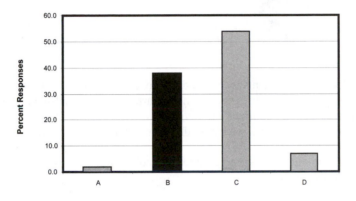

Comments: This question can be challenging for some students (62% selected incorrect answers). When discussing the different compounds in this question, we remind students that ionic bonding affects physical and chemical properties of ionic compounds.

8. Which structural isomer of C_5H_{12} will have the highest boiling point?

A) $CH_3CH_2CH_2CH_2CH_3$ (*n*-pentane)
B) $CH_3CH_2CH(CH_3)_2$ (2-methylbutane)
C) $(CH_3)_4C$ (2,2-dimethylpropane)
D) The boiling point is the same for all structural isomers.

Correct Answer: A
Level of Difficulty: Moderate
Keyword(s): Boiling points of structural isomers

Note: Histogram and comments for this question appear on the next page.

Histogram:

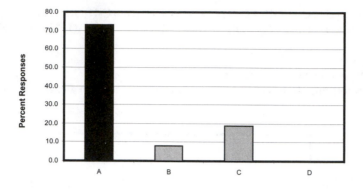

Comments: About 73% of the students recognized the difference in polarizabilities of the branched isomers compared to the straight chain hydrocarbons and the subsequent effect on boiling points. This was a good opportunity to review and reinforce the concept of constitutional isomers as students struggle with this concept both in general and organic chemistry.

9. Predict which liquid will have the <u>strongest</u> intermolecular forces of attraction (neglect the small differences in molar masses).

A) $CH_3COCH_2CH_2CH_3$ (molar mass = 86 g/mol)
B) $CH_3CH_2CH_2CH_2CH_2OH$ (molar mass = 88 g/mol)
C) $CH_3CH_2CH_2CH_2CH_2CH_3$ (molar mass = 86 g/mol)
D) $HOH_2C-CH=CH-CH_2OH$ (molar mass = 88 g/mol)

Correct Answer: D
Level of Difficulty: Easy
Keyword(s): Intermolecular forces
Comments: This appears to be an easy question as 90% answered correctly.

10. Predict which liquid will have the highest vapor pressure (neglect the small differences in molar masses).

A) $CH_3COCH_2CH_2CH_3$ (molar mass = 86 g/mol)
B) $CH_3CH_2CH_2CH_2CH_2OH$ (molar mass = 88 g/mol)
C) $CH_3CH_2CH_2CH_2CH_2CH_3$ (molar mass = 86 g/mol)
D) $HOH_2C-CH=CH-CH_2OH$ (molar mass = 88 g/mol)

Correct Answer: C
Level of Difficulty: Moderate/Challenging
Keyword(s): Intermolecular forces and vapor pressure
Comments: This question was challenging; less that 50% answered correctly. We use every opportunity we get to reinforce the inverse relationship between the strength of intermolecular forces and vapor pressure.

11. Consider an interior atom in the simple cubic crystal lattice. What is the maximum number of unit cells that share this atom in the three-dimensional crystal lattice?

A) 2   B) 4   C) 6   D) 8

Correct Answer: D
Level of Difficulty: Moderate/Challenging
Keyword(s): Simple cubic crystal lattice
Histogram:

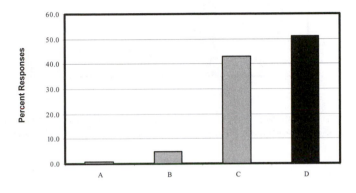

Comments: Our students had a difficult time visualizing crystal lattices and this histogram confirms this observation. We developed transparent models of simple, body-centered, and face-centered cubic unit cells that now produce responses of 94% or higher to questions dealing with cubic unit cells.

12. The number of atoms per unit cell in the body-centered cubic lattice is

A) 1   B) 2   C) 3   D) 4

Correct Answer: B
Level of Difficulty: Moderate
Keyword(s): Body-centered cubic unit cell
Comments: Some students struggle with this concept in class and develop more confidence after they work with models in the laboratory.

13. Select the correct statement pertaining to the properties of liquids.

     A) The meniscus in mercury is convex because the adhesive forces are stronger than the cohesive forces.
     B) Viscosity of a pure liquid decreases with increasing temperature.
     C) The surface tension of water is greater than that of mercury.
     D) Water rises up a glass capillary tube because the cohesive forces are stronger than the adhesive forces.

Correct Answer: B

Level of Difficulty: Moderate/Challenging
Keyword(s): Properties of liquids
Histogram:

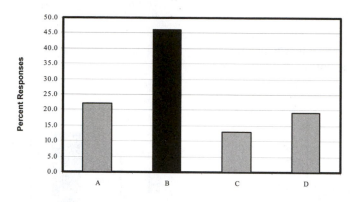

Comments: As the histogram shows, less than 50% answered correctly on the hour exam and there is a distribution of responses. Perhaps we do not spend enough time discussing these properties of liquids in class.

CHAPTER 20 Entropy and Free Energy

1. Recall the First Law of Thermodynamics. Which statements are true?

> I) The total energy of the universe is constant.
> II) $\Delta E_{univ} = \Delta E_{sys} + \Delta E_{surr} = 0$
> III) $\Delta E_{sys} = q + w$

> A) I and II   B) I and III   C) II and III   D) I, II and III

Correct Answer: D
Level of Difficulty: Easy/Moderate
Keyword(s): First Law of Thermodynamics
Histogram:

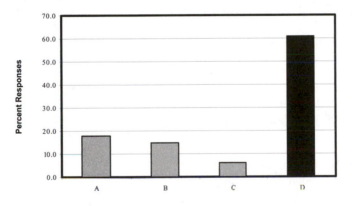

Comments: This histogram shows the results for individual participation prior to peer collaboration. The histogram was not displayed and the students were allowed to work with peers and click in a second time. Percent responses to distractors (A) and (B) decreased to 7% each while 20% selected distractor (C); the percent of correct responses decreased from 61% to 60%. The first law was discussed in the first semester course, and retention was tested prior to the introduction of entropy and free energy in the second semester.

2. Select the species in each pair with the higher S^0 value:

> I) $C_{graphite}(s)$ or $C_{diamond}(s)$ II) $CO(g)$ or $CO_2(g)$

> A) $C_{graphite}(s)$, $CO(g)$ B) $C_{graphite}(s)$, $CO_2(g)$
> C) $C_{diamond}(s)$, $CO(g)$ D) $C_{diamond}(s)$, $CO_2(g)$

Correct Answer: B
Level of Difficulty: Moderate
Keyword(s): Standard molar entropy, relative S^0 values

Histogram:

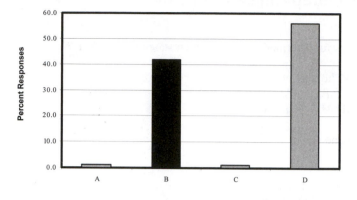

Comments: The histogram shows that 98% of the students recognized the higher S^0 value for CO_2 compared to CO and only 43% selected graphite as the correct answer. We used this opportunity to review the structures of diamond and graphite to explain the difference in relative S^0 values.

3. Which statement about standard molar entropy is correct?

> A) S^0 for $Br_2(g) < S^0$ for $Br_2(l)$
> B) S^0 for pure $O_2(g) > S^0$ for $O_2(g)$ dissolved in water
> C) S^0 for $KCl(s) > S^0$ for $KCl(aq)$
> D) S^0 for $N_2O_4(g) < S^0$ for $NO_2(g)$

Correct Answer: B
Level of Difficulty: Moderate/Challenging
Keyword(s): Relative standard molar entropies
Histogram:

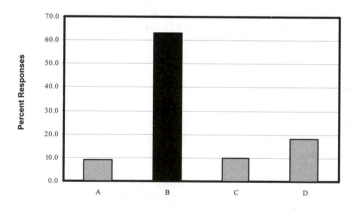

Comments: This question was used to conclude the discussion of relative S^0 values. About 63% responded correctly while others may need more time to reflect on the increase in entropy accompanying (i) increased molecular complexity, and (ii) dissolution of most ionic compounds.

4. Calculate ΔS^0 for the formation of 1 mol of HI(g) from its elements given the following information: $S^0[H_2(g)] = 131$ J/mol·K, $S^0[I_2(s)] = 116$ J/mol·K, $S^0[HI(g)] = 206$ J/mol·K.

A) 82 J/K   B) 165 J/K   C) 247 J/K   D) 329 J/K

Correct Answer: A
Level of Difficulty: Easy/Moderate
Keyword(s): Calculation of ΔS^0 using S^0 values
Histogram:

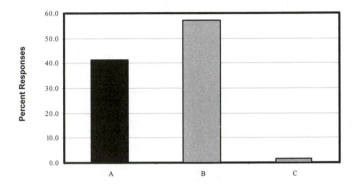

Comments: Our experience shows that students have a difficult time writing equations using fractional coefficients because we emphasize the use of smallest integer coefficients when balancing equations. The question clearly refers to the formation of 1 mole of HI(g); yet, 42% selected the distractor corresponding to the use of the equation $H_2(g) + I_2(g) \rightarrow 2HI(g)$.

5. Predict the sign of ΔS_{sys} for the reaction $N_2(g) + 3H_2(g) \rightarrow 2NH_3(g)$.

A) $\Delta S_{sys} = 0$ B) $\Delta S_{sys} > 0$
C) $\Delta S_{sys} < 0$ D) There is insufficient information to make a prediction.

Correct Answer: C
Level of Difficulty: Easy/Moderate
Keyword(s): Predicting ΔS_{sys}
Comments: Many of our students grasp this concept quite easily and 79% responded correctly.

6. For which reaction is $\Delta S_{sys} > 0$?

A) $CH_4(g) + 2O_2(g) \rightarrow CO_2(g) + 2H_2O(l)$
B) $NH_3(g) + HCl(g) \rightarrow NH_4Cl(s)$
C) $2NH_4NO_3(s) \rightarrow 2N_2(g) + O_2(g) + 4H_2O(g)$
D) $HCl(aq) + NaOH(aq) \rightarrow NaCl(aq) + H_2O(l)$

Correct Answer: C
Level of Difficulty: Easy/Moderate
Keyword(s): Predicting ΔS_{sys}

Comments: This question was posed on the hour exam and 84% answered correctly confirming our observation that students grasped this concept based on performance on clicker questions.

7. The normal boiling point of the element rubidium is 688°C. The following equilibrium is important in rubidium vapor at that temperature: $2 \text{ Rb}(g) \rightleftharpoons \text{Rb}_2(g)$. Predict the signs of ΔH^0 and ΔS^0 for this reaction.

A) ΔH^0 is negative, ΔS^0 is positive   B) ΔH^0 is negative, ΔS^0 is negative
C) ΔH^0 is positive, ΔS^0 is positive   D) ΔH^0 is positive, ΔS^0 is negative

Correct Answer: B
Level of Difficulty: Easy/Moderate
Keyword(s): Predicting signs of ΔH^0 and ΔS^0
Histogram:

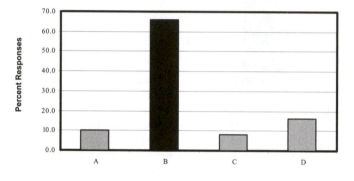

Comments: Based on our experience, the challenge was predicting the sign of ΔH^0 and not that of ΔS^0. About 24% did not recognize the release of energy accompanying the formation of a bond in the dimer. The misconception about energy changes in bond breaking and bond formation is also observed in first-semester organic chemistry as well as in many biology courses.

8. In a spontaneous process, there is an increase in the entropy of the universe. If $\Delta G = -T \Delta S_{\text{univ}}$, what is the sign of ΔG for a spontaneous process?

A) $\Delta G > 0$ B) $\Delta G = 0$ C) $\Delta G < 0$

Correct Answer: C
Level of Difficulty: Easy
Keyword(s): Sign of ΔG and spontaneity
Comments: This should be an easy concept (96% answered correctly) but it is better to engage the students in making this connection ("learning by doing") rather than providing the information. The use of clickers over several semesters has taught us not to make any assumptions about students' prior knowledge. This led us to develop simple questions like this one to facilitate student participation in the learning process.

9. For a reaction that is always non-spontaneous [reactant(s) favored], what are the signs of ΔH_{sys} and ΔS_{sys}?

A) $\Delta H_{sys} < 0$, $\Delta S_{sys} < 0$

B) $\Delta H_{sys} > 0$, $\Delta S_{sys} > 0$

C) $\Delta H_{sys} < 0$, $\Delta S_{sys} > 0$

D) $\Delta H_{sys} > 0$, $\Delta S_{sys} < 0$

Correct Answer: D
Level of Difficulty: Moderate/Challenging
Keyword(s): Effect of temperature on spontaneity
Histogram:

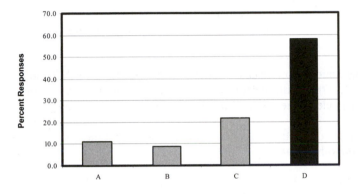

Comments: Some students struggle with the conceptual use of the Gibbs equation $\Delta G = \Delta H - T\Delta S$ to predict temperature effect on spontaneity of reactions. Only 58% selected the correct answer.

10. For which combination of signs will a reaction always be spontaneous?

	ΔS_{sys}	ΔH_{sys}
A)	−	−
B)	+	+
C)	−	+
D)	+	−

Correct Answer: D
Level of Difficulty: Moderate
Keyword(s): Effect of temperature on spontaneity
Comments: This question tests the same content (effect of T on spontaneity) as the previous one but the information is presented differently. About 78% selected the correct answer.

11. For which combination of signs will a reaction always be non-spontaneous?

	ΔS_{sys}	ΔH_{sys}
A)	–	–
B)	+	+
C)	–	+
D)	+	–

Correct Answer: C
Level of Difficulty: Moderate
Keyword(s): Effect of temperature on spontaneity
Comments: This question follows quite nicely from the previous one and the students participated very enthusiastically with 96% responding correctly. It is clear that they really enjoy doing well on clicker questions.

12. Hold the rubber band a short distance from your lips. Quickly stretch it and press it against your lips *carefully* (don't hurt those delicate lips). Do you experience a warming or cooling sensation? Carefully release the rubber band and experience the sensation.

Is stretching a spontaneous or a non-spontaneous process? Select the best answer for what happens when you stretch the rubber band.

A) $\Delta G^0 < 0, \Delta H^0 > 0$ B) $\Delta G^0 < 0, \Delta H^0 < 0$
C) $\Delta G^0 > 0, \Delta H^0 > 0$ D) $\Delta G^0 > 0, \Delta H^0 < 0$

Correct Answer: D
Level of Difficulty: Moderate/Challenging
Keyword(s): Experiencing the Gibbs equation
Histogram:

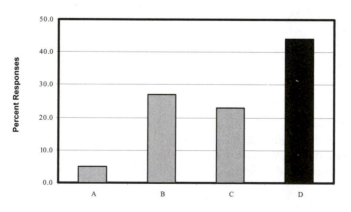

Comments: Students enjoy participating in this activity. The instrument is probably not very sensitive and may bias the results observed. There was quite a distribution of responses with only 44% answering correctly. We enjoy discussing the solution to this question as students keep repeating the experiment. Stretching is a non-spontaneous process ($\Delta G^0 > 0$) and a warming (ΔH^0

< 0) sensation is experienced. Use the Gibbs equation to show that $\Delta S^0 < 0$; this is explained by the increasing alignment of the polymer molecules compared to the randomness in the relaxed rubber band.

13. In the rubber band experiment (lecture demonstration), what happens when you allow the rubber band to relax? Select the correct signs for ΔG, ΔH, and ΔS.

	ΔG	ΔH	ΔS
A)	−	+	+
B)	−	−	+
C)	+	+	−
D)	+	−	−

Correct Answer: A
Level of Difficulty: Moderate/Challenging
Keyword(s): Using the Gibbs equation
Histogram:

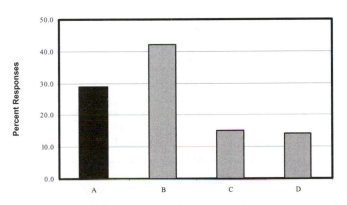

Comments: Only 29.2% answered correctly when this question was posed on the hour exam about 2 weeks after the experiment was performed in class.

14. Which statement is true about the equilibrium constant K for a reaction if ΔG^0 for that reaction is less than zero?

A) K = 0   B) K = 1   C) K>>1   D) K<<1

Correct Answer: C
Level of Difficulty: Easy/Moderate
Keyword(s): Relationship between ΔG^0 and K
Comments: This question was posed after the mathematical relationship between ΔG^0 and K was presented, and 82% responded correctly.

CHAPTER 21 Chemical Kinetics

1. If at a certain instant during the course of the reaction $2NO_2(g) + F_2(g) \rightarrow 2NO_2F(g)$, the rate of formation of the product were 0.16 mol/L·s, how fast would the F_2 be disappearing (also in mol/L·s) at that same instant in time?

A) 0.16 mol/L·s

B) 0.080 mol/L·s

C) 0.32 mol/L·s

D) 0.48 mol/L·s

Correct Answer: B
Level of Difficulty: Easy/Moderate
Keyword(s): Convention for reaction rate
Histogram:

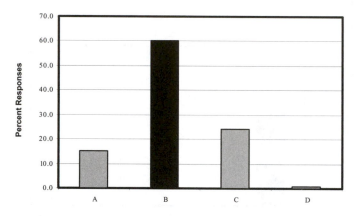

Comments: This question can be answered without any calculations by comparing the rate of disappearance of F_2 relative to the rate of appearance of the product based on stoichiometric ratios. Only 60% answered correctly confirming our earlier findings that many students are not comfortable with basic mathematical concepts. The mathematical description of reaction rate in terms of rates of disappearance of reactants and appearance of products was presented prior to this question. Yet, students could not use the algorithmic approach successfully, even with peer collaboration, in the brief time (1–2 minutes) allowed for conceptual questions.

2. The rate expression for the reaction between nitrogen dioxide and fluorine is determined experimentally and found to be rate = k [NO_2] [F_2]. If concentrations are expressed in moles per liter, what are the units of k, the rate constant?

A) s^{-1}

B) $mol \, L^{-1} s^{-1}$

C) $mol \, L^{-1} s$

D) $L \, mol^{-1} s^{-1}$

Correct Answer: D
Level of Difficulty: Easy/Moderate
Keyword(s): Units for rate constants

Comments: It is a good idea to use an example in class for obtaining the units of rate constants from the rate expression using dimensional analysis. Some students do not like the simple mathematical approach and tend to memorize units for rate constants.

3. A reaction obeys the rate law: rate = k $[X]^2[Y]$. If concentrations are expressed in mol/L and time in seconds, what are the units of k?

A) $L^3 mol^{-3} s^{-1}$ B) $L\ mol^{-1} s^{-1}$ C) $L^2 mol^{-2} s^{-1}$ D) $mol^2 L^{-2} s^{-1}$

Correct Answer: C
Level of Difficulty: Easy/Moderate
Keyword(s): Units for rate constants
Histogram:

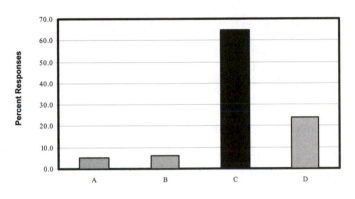

Comments: As mentioned in the previous example, some students continue to struggle with basic math skills throughout the course and do not gain confidence even with review and reinforcement. Even with peer collaboration, only 65% selected the correct answer.

4. Gallium-65, a radioactive isotope of gallium, decays by first-order kinetics. The half-life of this isotope is 15.2 minutes. How long would it take for ⅞ of a sample of this isotope to decay?

A) 15.2 min   B) 30.4 min   C) 45.6 min   D) 48.0 min

Correct Answer: C
Level of Difficulty: Easy/Moderate
Keyword(s): Half-life
Comments: This question could be relatively easy if students grasped the meaning of half-life and recognized that one-eighth of the sample remains after 3 half-lives. A similar question was given on a previous hour exam and less than half of the class selected the correct answer. Perhaps the students do not clearly distinguish between the amount that decays and the amount that remains at a certain point in time.

5. The rate constant, k, for the reaction: 2 NOBr → 2 NO + Br_2 is 0.80 $M^{-1}s^{-1}$. Which variables obtained for this reaction will produce a straight line?

A) log [NOBr] vs time   B) [NOBr] vs time
C) $[NOBr]^2$ vs time   D) $^1/_{[NOBr]}$ vs time

Correct Answer: D
Level of Difficulty: Easy/Moderate
Keyword(s): Integrated rate laws
Comments: This question links the use of the integrated rate laws with graphing variables to obtain linear plots. The challenging piece is using the units of the rate constant to determine the order of the reaction. This question helps students to develop and enhance analytical and critical thinking skills.

6. Consider the following hypothetical reaction: A + 2B → E

The mechanism for this reaction is:
(1)   A + B → C   (slow)
(2)   B + C → D   (fast)
(3)   D → E   (fast)

The rate law consistent with this mechanism is:

A) Rate = k [A] [B]   B) Rate = k $[A]^2$ [B]
C) Rate = k $[A]^2$ D) Rate = k [A] $[B]^2$

Correct Answer: A
Level of Difficulty: Easy/Moderate
Keyword(s): Reaction Mechanisms
Histogram:

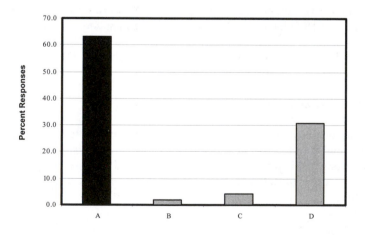

Comments: Understanding reaction mechanisms has always been challenging for our students. This mechanism was relatively simple with the first step being the slow, rate-determining step; yet, only 63% responded correctly even with peer collaboration. This does not surprise us as students were given little time to absorb the concept. Drill and practice is a very essential part of the learning process, and we use electronic homework that the students appreciate immensely.

7. Use the rate laws given below to determine which reaction is most likely to occur in a single step.

A) $2\ NO_2(g) + F_2(g) \rightarrow 2\ NO_2F(g)$ Rate $= k\ [NO_2]\ [F_2]$

B) $H_2(g) + Br_2(g) \rightarrow 2\ HBr(g)$ Rate $= k\ [H_2]\ [Br_2]^{1/2}$

C) $NO(g) + O_2(g) \rightarrow NO_2(g) + O(g)$ Rate $= k\ [NO]\ [O_2]$

D) $NO_2(g) + CO(g) \rightarrow NO(g) + CO_2(g)$ Rate $= k\ [NO_2]^2$

Correct Answer: C
Level of Difficulty: Easy/Moderate/Challenging
Keyword(s): Reaction mechanisms
Comments: This question tests understanding of elementary steps and use of the slow, rate-determining step to derive the rate expression. The correct answer appears to be a giveaway even for students who continue to cling to the misconception that the rate expression is derived from the balanced equation. Unfortunately, many students did not select the correct answer either as a conceptual question in class or on the exam.

8. Which statement is incorrect?

A) A catalyst provides an alternate mechanism for a reaction.
B) A catalyst is regenerated in a reaction.
C) A reaction involving a catalyst yields more product.
D) A catalyst speeds up the forward and reverse reactions.

Correct Answer: C
Level of Difficulty: Moderate
Keyword(s): Catalysts and catalysis

Note: Histogram and comments for this question appear on the next page.

Histogram:

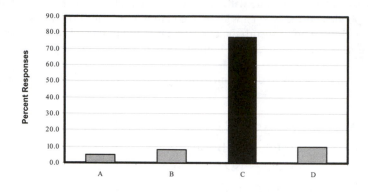

Comments: Students hear about catalysts in biology classes and have a decent grasp of the concept from a high school chemistry course. The correct answer was selected by 77% of the students just after the concept was presented in lecture. A very effective and colorful lecture demonstration that illustrates the formation of an intermediate and regeneration of the catalyst is the oxidation of tartaric acid (using Rochelle's salt) by H_2O_2 catalyzed by $CoCl_2$.

9. Does a catalyst increase reaction rate by the same means as a rise in temperature does?

A) Yes   B) No

Correct Answer: B
Level of Difficulty: Easy/Moderate
Keyword(s): Effect of catalyst on reaction rate compared to temperature effect
Histogram:

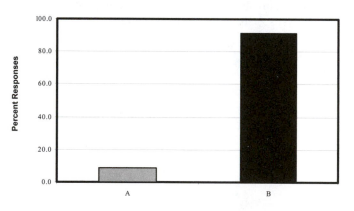

Comments: We do not usually use yes/no questions. However, this is an excellent question to engage the students and assess their understanding of the differences between the effects of a catalyst and temperature on reaction rates. We were very pleased to observe the students engage in thoughtful discussions when this question was presented and 91% responded correctly.

10. A catalytic mechanism proposed for the depletion of ozone by chlorofluorocarbons in the stratosphere is:

$$Cl + O_3 \rightarrow ClO + O_2$$
$$ClO + O \rightarrow Cl + O_2$$

Which statements about this mechanism are true?

> I) O_2 is a reactant.
> II) ClO is an intermediate.
> III) O_3 is decomposed to O_2 and O in the overall reaction.
> IV) Cl is the catalyst.

> A) I and II   B) III and IV   C) I and III   D) II and IV

Correct Answer: D
Level of Difficulty: Easy/Moderate
Keyword(s): Reaction mechanisms, intermediates, and catalysts
Comments: Students generally grasp the concepts about intermediates and catalysts quickly. However, we have found the need to review and reinforce these concepts in the sophomore organic chemistry courses.

11. Select the statement that is consistent with the potential energy diagram shown below.

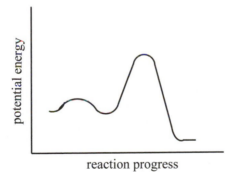

> A) The reaction is endothermic.
> B) Step 1 is the slow, rate-determining step.
> C) There are three intermediate species in the reaction.
> D) There are two transition states during the reaction.

Correct Answer: D
Level of Difficulty: Moderate
Keyword(s): Reaction mechanisms and potential energy profiles
Comments: We have found that our students are often challenged by the discussion of reaction mechanisms. A potential energy diagram can be an effective tool especially for visual learners. In

addition, this question reinforces concepts learned earlier in the course such as energy profiles for exothermic and endothermic reactions. This is also a very useful concept in organic chemistry courses.

CHAPTER 22 Chemical Equilibrium

1. Consider the equilibrium dissociation of $N_2O_4(g)$ to $NO_2(g)$. Use your knowledge of Lewis structures and bond energies (bond breaking vs. bond making) to predict whether the reaction $N_2O_4(g) \rightleftharpoons 2\,NO_2(g)$ is

A) exothermic   or   B) endothermic.

Correct Answer: B
Level of Difficulty: Easy/Moderate
Keyword(s): Introduction to chemical equilibrium
Comments: This reaction is the basis of an excellent lecture demonstration. We use 3 sealed tubes containing the same intensity (visibly light brown) of NO_2 gas at room temperature. One tube is maintained at room temperature, the second is cooled in dry ice, and the third is placed in hot water. After a short period of time, the three tubes are placed on the overhead projector. The low temperature tube is almost colorless while the high temperature tube is intensely brown compared to the room temperature tube. Students learn that NO_2 is a brown gas and N_2O_4 is colorless. Misconceptions about bond breaking and bond formation can be addressed. About 76% answered correctly.

2. Select the equilibrium expression for the homogeneous gas phase reaction:
$2SO_3(g) \rightleftharpoons 2SO_2(g) + O_2(g)$

A) $K_c = [SO_3] / ([SO_2][O_2])$ B) $K_c = [SO_3]^2 / ([SO_2]^2[O_2])$
C) $K_c = [SO_2] [O_2] / [SO_3]$ D) $K_c = [SO_2]^2 [O_2] / [SO_3]^2$

Correct Answer: D
Level of Difficulty: Easy
Keyword(s): Equilibrium expression
Comments: Students grasp this concept quite easily and as we expected 92% answered correctly.

3. Calculate K_c for the equilibrium $N_2(g) + 3H_2(g) \rightleftharpoons 2NH_3(g)$. The equilibrium amounts in a 2-L flask at 400 K are 0.02 mol N_2, 0.02 mol H_2, and 0.04 mol NH_3.

A) 2×10^2 B) 4×10^2 C) 1×10^4 D) 4×10^4

Correct Answer: D
Level of Difficulty: Moderate/Challenging
Keyword(s): Calculation of K_c

Note: Histogram and comments for this question appear on the next page.

Histogram:

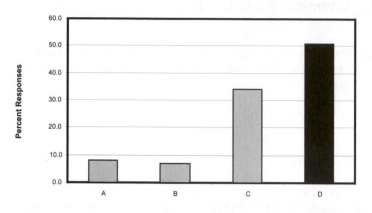

Comments: Students are comfortable with writing the equilibrium expression; however, only 51% got the correct answer in class. The relatively poor performance may be due to lack of confidence using exponents without a calculator.

4. The value of K_c for the reaction $H_2(g) \rightleftharpoons 2H(g)$ is 1.2×10^{-42} at 500 K. Select the correct statement about this system at equilibrium.

A) $[H_2] = [H]$
C) $[H] \gg [H_2]$

B) $[H_2] \gg [H]$
D) There is insufficient information to select a correct statement.

Correct Answer: B
Level of Difficulty: Moderate
Keyword(s): Magnitude of the equilibrium constant
Comments: This question helps students to conceptualize the relative concentrations of reactants and products at equilibrium as reflected by the magnitude of K_c. Some students need more examples to gain confidence using this concept. Using molecular views to convey this concept can be effective especially for visual learners.

5. At a particular temperature, $K_c = 25$ for the reaction: $2 NO(g) + 2 H_2(g) \rightleftharpoons N_2(g) + 2 H_2O(g)$

What is K_c for the reaction $NO(g) + H_2(g) \rightleftharpoons \frac{1}{2} N_2(g) + H_2O(g)$ at the same temperature?

A) 625   B) 50   C) 25   D) 5

Correct Answer: D
Level of Difficulty: Easy/Moderate
Keyword(s): Variations in the form of K_c

Histogram:

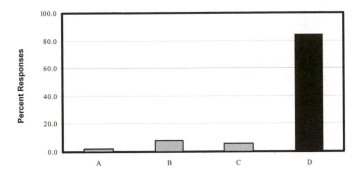

Comments: This question requires an algorithmic approach that students are generally comfortable with; this observation was confirmed as 84% responded correctly.

6. Phosgene, $COCl_2$, decomposes as shown: $COCl_2 (g) \rightleftharpoons CO(g) + Cl_2(g)$. K_c for the reaction is 8.3×10^{-4} at 360°C. Select the correct statement when 5.00 mol of $COCl_2$ decomposes at 360°C in a 10.0-L flask.

> A) The system is at equilibrium.
> B) The reverse reaction will predominate until equilibrium is established.
> C) The forward reaction will predominate until equilibrium is established.

Correct Answer: C
Level of Difficulty: Easy/Moderate
Keyword(s): Comparing Q_c and K_c to predict direction of reaction
Histogram:

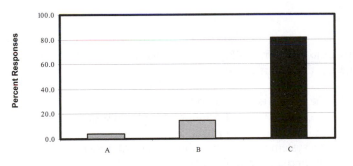

Comments: This question requires comparison of Q_c and K_c to predict the direction in which reaction proceeds until equilibrium is established. In this example, students will learn that the reaction must proceed to the right as we start with only reactant and no products. About 81% selected the correct answer.

7. For the water gas shift reaction, $CO(g) + H_2O(g) \rightleftharpoons CO_2(g) + H_2(g)$, $K_c = 1.00$ at 1100 K. The following amounts are brought together in a 1.0-L container and allowed to react at 1100 K: 1.0 mol of CO, 1.0 mol of H_2O, 2.0 mol of CO_2 and 2.00 mol of H_2. Select the correct statement about this system.

        A) The system is at equilibrium.
        B) The reverse reaction will predominate until equilibrium is established.
        C) The forward reaction will predominate until equilibrium is established.

Correct Answer: B
Level of Difficulty: Moderate
Keyword(s): Comparing Q_c and K_c to predict direction of reaction
Histogram:

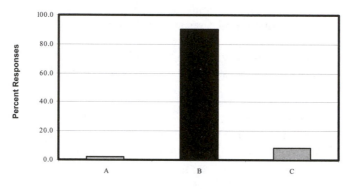

Comments: This question requires calculation and comparison of Q_c and K_c to predict the direction in which reaction proceeds until equilibrium is established. This example shows that the calculation is essential when all reactants and products are present. We were pleased that 90% responded correctly.

8. Consider the homogeneous gas phase reaction $N_2O_4(g) \rightleftharpoons 2\ NO_2(g)$; $K_c = 0.211$ at 100°C. At some point during the reaction, $[N_2O_4] = 0.12$ M and $[NO_2] = 0.55$ M. Select the correct statement about the reaction at this point.

        A) The system is at equilibrium.
        B) The system is not at equilibrium and will proceed towards the product until equilibrium is established.
        C) The system is not at equilibrium and will proceed towards the reactant until equilibrium is established.
        D) There is insufficient information to determine whether the system is at equilibrium or not.

Correct Answer: C
Level of Difficulty: Moderate
Keyword(s): Comparing Q_c and K_c to predict direction of equilibrium

Comments: This question requires calculation and comparison of Q_c and K_c to predict the direction in which reaction proceeds until equilibrium is established. This example shows that the calculation is essential when the reactant and product are present. The correct answer was obtained by 77% of the students.

9. For which of these equilibria is $K_P \neq K_c$?

I) $2HgO(s) \rightleftharpoons 2Hg(l) + O_2(g)$
II) $SO_2Cl_2(g) \rightleftharpoons SO_2(g) + Cl_2(g)$
III) $N_2(g) + O_2(g) \rightleftharpoons 2\,NO(g)$

A) I and II   B) I and III   C) II and III   D) I, II and III

Correct Answer: A
Level of Difficulty: Moderate
Keyword(s): Comparing K_P and K_c
Histogram:

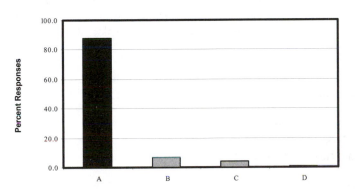

Comments: Students will learn that the change (Δn) in the number of moles of gaseous reactants and products determines the relationship between K_P and K_c. The correct answer was obtained by 88% of the students.

10. For the equilibrium $(NH_4)_2CO_3(s) \rightleftharpoons 2NH_3(g) + CO_2(g) + H_2O(g)$, $\Delta H^0 = +33$ kJ/mol. The reactant would be favored at equilibrium if

A) the volume of the container is decreased at constant T.
B) some $(NH_4)_2CO_3(s)$ is added.
C) the temperature is increased.
D) some ammonia is removed.

Correct Answer: A
Level of Difficulty: Challenging
Keyword(s): Le Châtelier's Principle

Histogram:

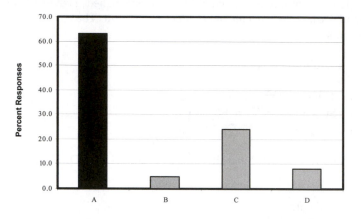

Comments: This question is thought provoking as it includes several factors that affect a system at equilibrium. The effect of volume-changes on the direction in which equilibrium shifts to offset the stress is one of the more challenging factors that students encounter. It was not surprising that only 63% answered correctly. It is important to take time to explain the effect of each factor used as a distractor.

11. Consider the equilibrium: I_2 (g) $\rightleftharpoons$ 2I (g). Predict the dependence of the equilibrium constant on temperature.

        A) There is no dependence of K_c on T.
        B) K_c increases as T increases.
        C) K_c decreases as T increases.

Correct Answer: B
Level of Difficulty: Moderate
Keyword(s): Effect of temperature on equilibrium constant; Le Châtelier's Principle
Histogram:

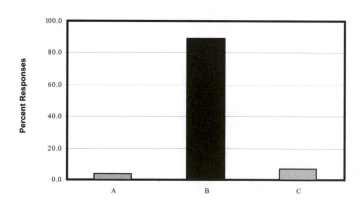

Comments: This question required students to recognize that the reaction involved breaking a bond and was therefore endothermic. This information was then used to predict the effect of temperature on K_c; 89% answered correctly.

12. Which change will have no effect on the position of equilibrium for the endothermic reaction $H_2(g) + I_2(g) \rightleftharpoons 2HI(g)$?

  A) Addition of $H_2(g)$ at constant temperature
  B) Removal of $HI(g)$ at constant temperature
  C) Doubling the volume of the container at constant temperature
  D) Decreasing the temperature

Correct Answer: C
Level of Difficulty: Moderate/Challenging
Keyword(s): Le Châtelier's Principle
Comments: This question can be used to conclude the discussion of the factors that affect the equilibrium position or to review for an exam.

CHAPTER 23 Acid-Base Equilibria

1. What are the major species present in an aqueous solution of 1.0 M HCl?

    I) HCl   II) H_2O III) H_3O^+ IV) Cl^- V) OH^-

        A) I and II   B) II, III, and IV
        C) I, II, III, and IV   D) I, II, III, IV, and V

Correct Answer: B
Level of Difficulty: Easy/Moderate
Keyword(s): Dissociation of strong acids
Histogram:

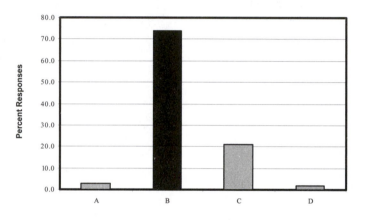

Comments: Acid-base chemistry has been one of the most challenging topics for both our students and first-year graduate teaching assistants. Students have a difficult time predicting products of these reactions and are challenged by the calculation of pH of these solutions. Our experience has shown that recognition of the predominant species in solution and visualization of the acid-base process makes the task a little less frustrating for students. In recent years, we have focused on developing materials to help with visualization and learned that student progress in understanding acid-base concepts is still very slow. Strong acids are less challenging, and 74% responded correctly to this question.

2. What are the major species present in an aqueous solution of 1.0 M CH_3COOH?

    I) CH_3COOH II) H_2O III) H_3O^+ IV) CH_3COO^- V) OH^-

        A) I and II   B) II, III, and IV
        C) I, II, III, and IV   D) I, II, III, IV, and V

Correct Answer: A

Level of Difficulty: Moderate/Challenging
Keyword(s): Dissociation of weak acids
Histogram:

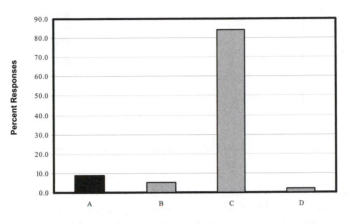

Comments: Only 9% correctly identified the major species in an aqueous solution of a weak acid. Students have a difficult time distinguishing between strong and weak acids. In addition, they are frequently not confident in defining major and minor species in solution. As the histogram shows, 84% selected the distractor that included water molecules, hydronium ions, and the conjugate base of the acid (as in the case of strong acids) although they recognized the presence of the weak acid molecules in solution. We are currently using both molecular/ionic views as well as graphical representations for the ionization of weak acids.

3. What are the major species present in solution when equal volumes of 0.10 M solutions of CH_3COOH and $NaOH$ are reacted?

A) CH_3COOH, $NaOH$, and H_2O
C) $NaC_2H_3O_2(s)$ and H_2O

B) H_3O^+, OH^- and H_2O
D) Na^+, CH_3COO^- and H_2O

Correct Answer: D
Level of Difficulty: Moderate/Challenging
Keyword(s): Neutralization reactions

Note: Histogram and comments for this question appear on the next page.

Histogram:

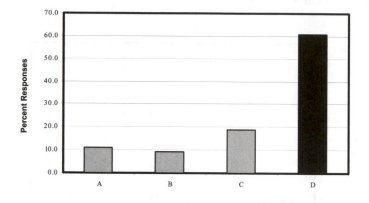

Comments: Neutralization reactions were introduced in the first semester and students have learned to write a balanced equation. Our goal is to encourage them to visualize the process and predict what ions and molecules are present after neutralization occurs, as well as the relative concentrations of these. 61% answered correctly and 19% selected distractor (C). The question was revised based on issues raised by the students in class. Equimolar solutions in the original question was replaced by 0.10 M solutions, and $NaC_2H_3O_2(s)$ replaced $NaC_2H_3O_2$.

4. Examine the K_a values for the acids shown below.

Acid	Name	K_a
HOBr	Hypobromous acid	2.3×10^{-9}
CH_3COOH	Acetic acid	1.8×10^{-5}
HCOOH	Formic acid	1.8×10^{-4}
(OH on benzene ring)	Phenol	1.0×10^{-10}

Consider 0.10 M solutions of each acid. Arrange these solutions in order of increasing $[H_3O^+]$.

    A) Acetic acid < Formic acid < Hypobromous acid < Phenol
    B) Formic acid < Acetic acid < Hypobromous acid < Phenol
    C) Phenol < Hypobromous acid < Acetic acid < Formic acid
    D) Phenol < Hypobromous acid < Formic acid < Acetic acid

Correct Answer: C
Level of Difficulty: Easy
Keyword(s): Acid dissociation constants, K_a values, and pH
Comments: We were pleased when 93% answered correctly confirming that students quickly grasp the meaning of the magnitude of K_a and its effect on $[H_3O^+]$ (and pH).

5. Which does **not** constitute a conjugate acid-base pair?

A) $H_3PO_4 / H_2PO_4^-$
B) $H_2PO_4^- / PO_4^{3-}$
C) HPO_4^{2-} / PO_4^{3-}
D) $H_2PO_4^- / HPO_4^{2-}$

Correct Answer: B
Level of Difficulty: Easy
Keyword(s): Conjugate acid-base pairs
Comments: Our students mastered this concept quickly and 94% responded correctly.

6. Predict the magnitude of the equilibrium constant for the reaction:
$NH_3(aq) + CH_3COOH(aq) \rightleftharpoons$

A) $K = 0$
B) $K = 1$
C) $K \ll 1$
D) $K \gg 1$

Correct Answer: D
Level of Difficulty: Challenging
Keyword(s): Predicting direction of acid-base equilibria
Histogram:

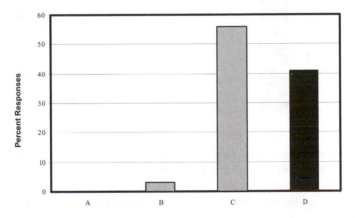

Comments: Only 41% responded correctly. This is a challenging concept and the question allows the instructor to review and reinforce concepts learned earlier. We slowly walk our students through the process of predicting the products based on proton transfer and comparing the relative acid strengths of CH_3COOH and NH_4^+. We also review the concept of multiple equilibria as we calculate the equilibrium constant for the reaction and reassure students that the qualitative prediction was correct.

7. 10.0 mL of 1.0 M HNO_2 is diluted to a final volume of 1.0 L. Select the incorrect statements comparing the solutions before and after dilution.

     I) Concentration of diluted HNO_2 solution is 0.10 M.
     II) $[H_3O^+]$ increases on dilution
     III) % ionization increases on dilution

   A) I and II   B) II and III   C) I and III   D) I, II and III

Correct Answer: A
Level of Difficulty: Challenging
Keyword(s): Dilution of weak acids and effect on $[H_3O^+]$ and percent ionization
Histogram:

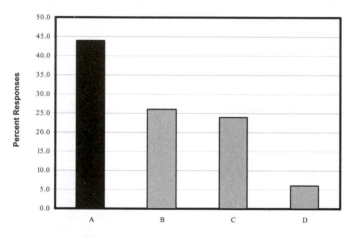

Comments: This is another challenging concept for our students; only 44% responded correctly even with peer collaboration. Review and reinforcement are highly recommended.

8. Predict the relative magnitudes of the successive acid dissociation (ionization) constants for H_3PO_4 (phosphoric acid).

    A) $K_{a1} > K_{a2} > K_{a3}$ B) $K_{a3} > K_{a2} > K_{a1}$
    C) $K_{a1} > K_{a2} < K_{a3}$ D) $K_{a1} < K_{a2} > K_{a3}$

Correct Answer: A
Level of Difficulty: Moderate
Keyword(s): Polyprotic acids and successive ionization constants

Note: Histogram and comments for this question appear on the next page.

Histogram:

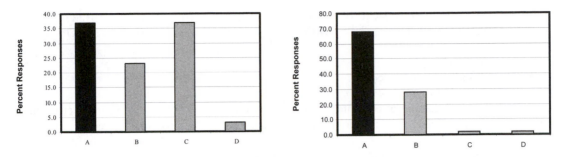

Comments: This question was designed to test the reading assignment. The histogram on the left represents individual student performance, followed by responses after peer collaboration (histogram on right). The histogram on the left was not displayed prior to collaboration with peers. These histograms support our philosophy to encourage peer collaboration; 68% responded correctly after interacting with their peers compared to 37% when working individually.0

9. What are the major species present in a 0.10 M solution of sodium cyanide, NaCN?

A) Na^+ and CN^-
C) Na^+, CN^-, H_3O^+ and OH^-

B) Na^+, CN^-, and H_2O
D) Na^+, CN^-, H_2O, H_3O^+ and OH^-

Correct Answer: B
Level of Difficulty: Moderate
Keyword(s): Acid-base properties of salt solutions
Histogram:

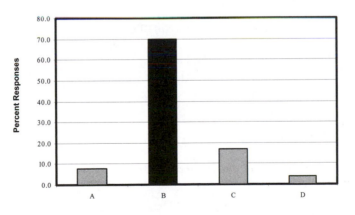

Comments: Predicting the acid-base properties of salt solutions is a major hurdle for our students. This question was answered correctly by 70% of the students.

10. Select the correct order of acid strengths.

A) $HClO > HBrO > HClO_2 > HNO_3$ B) $HBrO > HClO > HClO_2 > HNO_3$

C) $HNO_3 > HClO_2 > HClO > HBrO$ D) $HClO_2 > HNO_3 > HClO > HBrO$

Correct Answer: C
Level of Difficulty: Easy/Moderate
Keyword(s): Predicting relative acid strengths
Comments: This question was presented after a discussion of the effect of molecular structure on the acid strengths of the hydrohalic acids and oxyacids. About 84% responded correctly.

11. Arrange these 0.10 M solutions in order of increasing pH:

$$NaCl, NH_4Cl, HCl, CH_3COONa, KOH$$

A) $NH_4Cl, NaCl, CH_3COONa, HCl, KOH$
B) $KOH, NaCl, NH_4Cl, HCl, CH_3COONa$
C) $HCl, CH_3COONa, NaCl, NH_4Cl, KOH$
D) $HCl, NH_4Cl, NaCl, CH_3COONa, KOH$

Correct Answer: D
Level of Difficulty: Moderate
Keyword(s): Relative pH values of acids, bases, and salt solutions
Comments: The correct response (93%) was much higher than expected. A possible explanation is that we used $NaCl$, NH_4Cl, and CH_3COONa as examples when discussing the pH of salt solutions just before this question was presented.

12. Which of these 0.10 M solutions will have a pH lower than 7.00?

I) KCN   II) NH_4I III) NaH_2PO_4 IV) $Fe(NO_3)_3$

A) I and II   B) I and III
C) I, II and IV   D) II, III, and IV

Correct Answer: D
Level of Difficulty: Moderate
Keyword(s): pH of salt solutions

Note: Histogram and comments for this question appear on the next page.

Histogram:

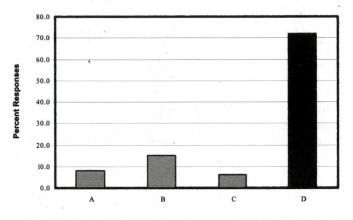

Comments: Review and reinforcement of the acid-base properties of salt solutions is highly recommended. About 72% of our students selected the correct answer. We also link this information to the experiments performed in the laboratory.

13. Select the correct statement.

A) Al^{3+} is a Brönsted-Lowry acid.   B) Al^{3+} is a Brönsted-Lowry base.
C) Al^{3+} is a Lewis acid.   D) Al^{3+} is a Lewis base.

Correct Answer: C
Level of Difficulty: Easy/Moderate
Keyword(s): Lewis acid-base concept
Comments: This question was answered correctly by 90% of our students. This is not a good representation of the hurdles that students face when applying the Lewis concept to other examples as evidenced on exams.

14. Select all the correct statements about an acid-base buffer.

I) It contains a strong acid and its conjugate base.
II) It contains a weak acid and its conjugate base.
III) It contains a strong base and its conjugate acid.
IV) It contains a weak base and its conjugate acid.
V) It resists changes in pH when small amounts of acid or base are added.

A) II and IV   B) I, III, and V   C) I, IV, and V   D) II, IV, and V

Correct Answer: D
Level of Difficulty: Easy/Moderate
Keyword(s): Buffers
Comments: This question was presented at the start of class to review material covered in the previous lecture and 83% answered correctly.

15. Which combinations will not serve as buffers when equal volumes of 0.20 M solutions are mixed?

I) $HClO_4/KClO_4$ II) HF/NaF III) NaH_2PO_4/Na_2HPO_4 IV) HNO_3/KNO_3

   A) II and III   B) III and IV   C) I and III   D) I and IV

Correct Answer: D
Level of Difficulty: Moderate
Keyword(s): Recognizing components of a buffer
Comments: Only 39% responded correctly. Our experience shows that students have a difficult time distinguishing between strong and weak acids. We encourage our students to recognize $HClO_4$, HNO_3, H_2SO_4, HCl, HBr, and HI as the six strong acids that they will encounter in our general chemistry course.

16. Which of these is **not** a good method for preparing a buffer? (HA is a weak acid)

   A) Equal volumes of 0.5 M HA and 0.5 M NaA
   B) Equal volumes of 1.0 M HA and 0.5 M NaOH
   C) Equal volumes of 1.0 M HCl and 0.5 M NaA
   D) Equal volumes of 0.5 M NH_3 and 0.5 M NH_4Cl

Correct Answer: C
Level of Difficulty: Moderate
Keyword(s): Preparation of a buffer (qualitative)
Comments: Our students were working on a lab experiment that involved the preparation of a 1:1 acetic acid-acetate buffer in three different ways. Perhaps this explains the rather high correct response (86%) to this question.

17. Which solution will have the highest buffer capacity?

   A) 1.0 M HF / 0.010 M NaF   B) 0.10 M H_3PO_4 / 0.10 M Na_3PO_4
   C) 0.10 M $NaHCO_3$ / 0.10 M Na_2CO_3 D) 1.0 M H_3PO_4 / 1.0 M Na_2HPO_4

Correct Answer: C
Level of Difficulty: Moderate
Keyword(s): Buffer capacity
Comments: Most students ruled out distractors (A) and (B); 57% selected the correct response, and 29% selected distractor (D), probably based on the relatively higher concentration but overlooking the absence of the conjugate acid-base relationship. Unfortunately, students refer to these examples as "tricky" questions.

18. A chemist needs to prepare a solution buffered at pH = 4.30 using benzoic acid (pK_a = 4.19) and its sodium salt. Calculate the ratio [HA] / [A^-].

   A) 0.11   B) 0.78   C) 0.96   D) 1.3

Correct Answer: B
Level of Difficulty: Moderate
Keyword(s): Preparation of a buffer (quantitative)
Histogram:

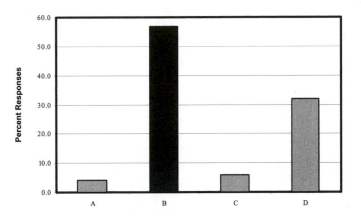

Comments: Using the Henderson-Hasselbalch equation with the data given in the question, students obtain a value of 1.3 for the ratio $[A^-] / [HA]$ and 32% selected distractor (D). The question asked for the ratio $[HA] / [A^-]$ which is 0.78 (reciprocal of 1.288) and 57% arrived at the correct answer.

19. Consider the titration of 50.0 mL of 0.200 M HNO_3 with 0.100 M NaOH solution. What volume of NaOH is required to reach the equivalence point in the titration?

    A) 25.0 mL   B) 50.0 mL   C) 1.00×10^2 mL   D) 1.50×10^2 mL

Correct Answer: C
Level of Difficulty: Easy
Keyword(s): Strong acid-strong base titration
Comments: This question tests retention of skills learned in the first semester and 89% arrived at the correct answer. Our experience confirms the benefits of reviewing these skills in preparation for the discussion on titration curves.

20. Consider the following acid-base titrations:

        I) 50 mL of 0.1 M HCl is titrated with 0.2 M KOH.
        II) 50 mL of 0.1 M CH_3COOH is titrated with 0.2 M KOH.

Which statement is not true?

    A) The initial pH of I is lower than that of II.
    B) The equivalence point for I and II will occur at the same volume of base.
    C) The equivalence point in I occurs at a lower pH than in II.
    D) There will be a buffer region in both I and II around the half-equivalence point.

Correct Answer: D
Level of Difficulty: Moderate
Keyword(s): Acid-base titration curves
Comments: This is a good question to compare and contrast the titrations of strong and weak acids with a strong base.

CHAPTER 24 Solubility and Complex-Ion Equilibria

1. Based on your knowledge of the solubility rules for ionic compounds in water, which salt is relatively the most soluble in water?

A) $CaCO_3$ B) $(NH_4)_3PO_4$ C) $BaSO_4$ D) $Mg(OH)_2$

Correct Answer: B
Level of Difficulty: Moderate
Keyword(s): Solubility rules
Histogram:

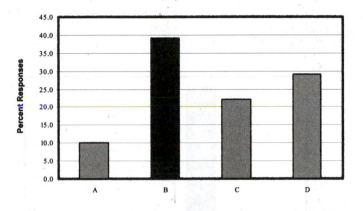

Comments: This question was presented before the review and discussion of the solubility rules to test prior knowledge. The histogram shows that 61% of the students could not identify the relatively higher solubility of the ammonium salt. We have learned that it is important to gauge prior knowledge and to review and reinforce concepts whenever the opportunity arises. This approach can improve learning and long-term retention of knowledge.

2. A 1-L beaker contains 250 mL of a saturated solution (in contact with undissolved solid) of a sparingly soluble salt, MX. Select the incorrect statement about this solution.

A) $[M^+] = [X^-]$
B) After adding 250 mL of pure distilled water and stirring the contents, some solid remains undissolved; $[M^+] = [X^-]$ in this solution.
C) For the solution in B), $[M^+]$ is the same as the original 250 mL of solution in the 1-L beaker.
D) For the solution in B), the number of M^+ ions is the same as in the original 250 mL of solution in the 1-L beaker.

Correct Answer: D
Level of Difficulty: Moderate/Challenging
Keyword(s): Saturated solutions of sparingly soluble salts

Comments: Our experience confirms that students are challenged by the solubility of sparingly soluble salts. Visualization of the process can be very effective in learning this concept. We usually draw cartoons on the overhead projector of two identical containers with different volumes of solution in contact with undissolved solid. We also use a very effective simulation about sparingly soluble salts that is available for public access at the PhET site maintained by the physics department at CU-Boulder.

3. Which equation best represents the relationship between K_{sp} and molar solubility (S) of silver chromate, Ag_2CrO_4?

$\quad\quad\quad$ A) $K_{sp} = S^2$ $\quad\quad\quad\quad\quad\quad$ B) $K_{sp} = 4S^3$

$\quad\quad\quad$ C) $K_{sp} = 27S^4$ $\quad\quad\quad\quad\quad$ D) $K_{sp} = 108S^5$

Correct Answer: B
Level of Difficulty: Easy/Moderate
Keyword(s): Molar solubility and K_{sp}
Histogram:

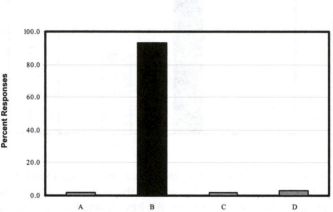

Comments: This question becomes relatively easy when students write down the equation for the solubility of the sparingly soluble and recognize that a molar solubility S implies $[CrO_4^{2-}] = S$ and $[Ag^+] = 2S$. In our class, 93% responded correctly.

4. For which of these sparingly soluble salts is K_{sp} related to the molar solubility (S) by the relationship $K_{sp} = 4S^3$?

$\quad\quad\quad\quad$ I) $Pb(IO_3)_2$ $\quad\quad\quad$ II) $BaSO_4$ $\quad\quad\quad$ III) Ag_2SO_4

$\quad\quad$ A) I and II $\quad\quad\quad$ B) I and III $\quad\quad\quad$ C) II and III $\quad\quad\quad$ D) I, II, and III

Correct Answer: B
Level of Difficulty: Easy/Moderate
Keyword(s): Molar solubility and K_{sp}

Histogram:

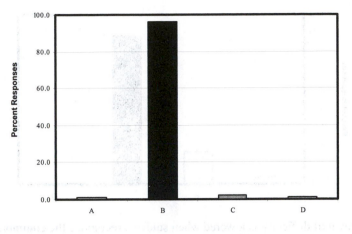

Comments: This question is similar to the previous one and 96% selected the correct answer. It helps to recognize that a 1:2 (ratio of ions) salt has the same relationship between S and K_{sp} as a 2:1 salt.

5. Which silver salt has the highest molar solubility in water at 25°C?

  A) Ag_2CrO_4 ($K_{sp} = 2.6 \times 10^{-12}$)   B) $AgCl$ ($K_{sp} = 1.8 \times 10^{-10}$)
  C) $AgBr$ ($K_{sp} = 5.0 \times 10^{-13}$)   D) AgI ($K_{sp} = 8.3 \times 10^{-17}$)

Correct Answer: A
Level of Difficulty: Moderate
Keyword(s): Relative solubilities of sparingly soluble salts
Comments: It is necessary for students to recognize that K_{sp} values can be used to compare relative solubilities only if the salts have the same ratio of ions. If they have different ratios of ions, then molar solubilities must be calculated and compared. In this example, the silver halides have the same ionic ratio and AgCl is more soluble than AgBr and AgI using K_{sp} values. A comparison of molar solubilities of AgCl and Ag_2CrO_4 shows that the latter is relatively more soluble.

6. Predict how the molar solubility of AgCl will change in a solution of 0.10 M NaCl compared to a saturated AgCl solution.

  A) Molar solubility will not change.
  B) Molar solubility will increase and more solid will dissolve.
  C) Molar solubility will decrease and some solid will precipitate out of solution.
  D) There is insufficient information to make a prediction.

Correct Answer: C
Level of Difficulty: Easy/Moderate
Keyword(s): Common ion effect on solubility of sparingly soluble salts

Histogram:

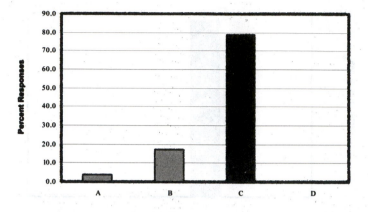

Comments: The level of difficulty is lowered when students recognize the common ion effect and apply Le Châtelier's principle correctly. About 79% of our students selected the correct answer.

7. Predict what will happen when the pH of a saturated $Mg(OH)_2$ solution is increased to 12.00. For a saturated $Mg(OH)_2$ solution, the pH = 11.03.

A) More $Mg(OH)_2$ will dissolve.
B) Some $Mg(OH)_2$ will precipitate out.
C) A pH change has no effect on the solubility of $Mg(OH)_2$.
D) There is insufficient information to make a prediction.

Correct Answer: B
Level of Difficulty: Moderate
Keyword(s): Effect of pH on solubility of sparingly soluble salts
Histogram:

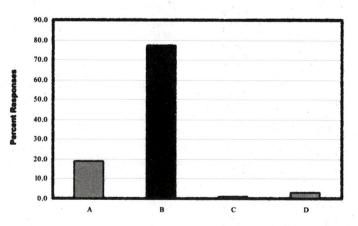

Comments: Students may be reminded about writing the correct balanced equation that relates the undissolved salt and its ions in solution. The presence of hydroxide ions in solution may provide a link to the effect of pH. About 77% responded correctly while 19% who selected distractor (A) misinterpreted the pH effect on solubility. This question also reviews and reinforces the pH concept.

8. You have a saturated aqueous solution of Ag_2CrO_4 at 25°C. Select all correct statements about this saturated solution assuming temperature remains constant.

> I) $[Ag^+] = 2[CrO_4^{2-}]$
> II) Molar solubility of Ag_2CrO_4 increases on dilution with water, assuming some solid remains undissolved.
> III) Molar solubility of Ag_2CrO_4 decreases when aqueous $AgNO_3$ solution is added.

A) I and II   B) I and III   C) II and III   D) I, II and III

Correct Answer: B
Level of Difficulty: Moderate
Keyword(s): Saturated solution of sparingly soluble salt, molar solubility and concentration of ions, common ion effect
Comments: This question may be used to summarize the discussion of sparingly soluble salts and factors that influence solubility of these salts.

9. Which of these actions will increase the solubility of $Cu(OH)_2$ in a saturated aqueous solution?

> I) A small amount of crystalline $Cu(NO_3)_2$ is added.
> II) The solution is buffered at pH = 2.00.
> III) Ammonia gas is bubbled through the solution.

A) I and II   B) II and III   C) I and III   D) I, II and III

Correct Answer: B
Level of Difficulty: Moderate/Challenging
Keyword(s): Factors that influence the solubility of sparingly soluble salts

Note: Histogram and comments for this question appear on the next page.

Histogram:

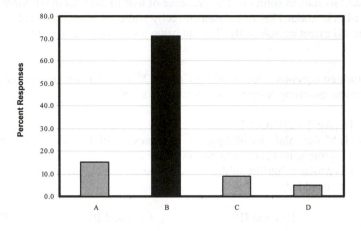

Comments: The correct answer was selected by 71% of the students. This question was presented after the discussion of complex ion formation on solubility of sparingly soluble salts. Our students do not remember the deep blue color of the $[Cu(NH_3)_4]^{2+}$ ion even though they produce it in the lab in several experiments. It might be a good idea to perform the lecture demonstration after the histogram is displayed.

10. Which statement is incorrect?

A) The molar solubility of AgI ($K_{sp} = 8.3 \times 10^{-17}$) is greater than that of AgCl ($K_{sp} = 1.8 \times 10^{-10}$) when saturated aqueous solutions are compared.

B) The molar solubility, S, of AgI in a saturated solution is related to its K_{sp} by the expression: $S = (K_{sp})^{1/2}$.

C) The molar solubility of AgI in 0.10 M $AgNO_3$ is lower than in pure water.

D) The molar solubility of AgI in NH_3 is greater than in pure water.

Correct Answer: A
Level of Difficulty: Moderate/Challenging
Keyword(s): Factors that influence the solubility of sparingly soluble salts
Comments: This question reviews several of the factors that affect solubility of sparingly soluble salts and provides a good opportunity to summarize the key points in the unit on solubility and complex-ion equilibria. Our students generally have a difficult time with both the qualitative and quantitative aspects of this unit.

CHAPTER 25 Transition Metals and Coordination Chemistry

1. What is the correct electron configuration for $_{24}Cr$?

A) $[_{18}Ar]\ 3d^4\ 4s^2$ B) $[_{18}Ar]\ 3d^5\ 4s^1$ C) $[_{18}Ar]\ 4s^2\ 4p^4$ D) $[_{18}Ar]\ 4s^1\ 4p^5$

Correct Answer: B
Level of Difficulty: Easy/Moderate
Keyword(s): Electron configurations of transition metals
Histogram:

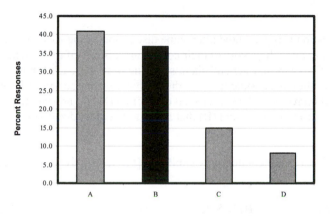

Comments: This question tested prior knowledge from first-semester chemistry. The histogram shows that the students remembered the filling of $3d$ orbitals in the first transition series but did not remember the exceptions with Cr (and Cu) due to the stability associated with half-filled (and filled) d-sublevels. About 37% remembered this detail while 41% selected distractor (A). Review and reinforcement of this information is appropriate as a prelude to the discussion of transition metal chemistry.

2. Which statement is <u>incorrect</u>?

         A) Many transition metals exhibit several different oxidation states.
         B) The 2+ oxidation state is common for many transition metals in the $3d$ series.
         C) The period 6 transition metals have much larger radii than the period 5 transition metals.
         D) In a given period, the atomic properties (like atomic radii, electronegativity, and first ionization energy) of the transition metals are more similar than those of the main group elements.

Correct Answer: C
Level of Difficulty: Moderate/Challenging
Keyword(s): Properties of transition metals

Histogram:

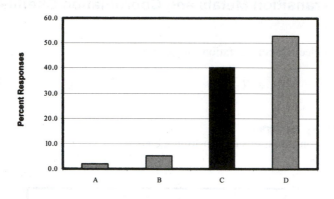

Comments: This question was presented after some discussion of the various oxidation states of transition metals that explain the colors of their compounds. Some of the statements refer to concepts covered in first-semester general chemistry. The best approach for students at this stage would be to eliminate the correct statements as the lanthanide contraction had not been discussed. Only 40% responded correctly and 53% selected distractor (D) that appears to be consistent with confidence levels for distractors (A) and (B) that relate to the class discussion on variable oxidation states.

3. Which salt will have a molar conductivity similar to $[Pt(NH_3)_4]Cl_2$? Assume that each aqueous solution is 0.10 M.

        A) $AlCl_3$ B) K_2SO_4 C) $CuSO_4$ D) $Al_2(SO_4)_3$

Correct Answer: B
Level of Difficulty: Easy/Moderate
Keyword(s): Conductivity of solutions of coordination compounds; understanding complex ions
Histogram:

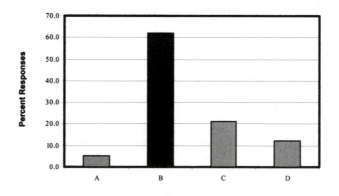

Comments: Our experience shows that students are challenged by the concept of electrolytes. This question aims at reinforcing the electrolyte concept using conductivity and providing some comparison with familiar salt solutions that do not contain complex ions. We use this opportunity to discuss the meaning of the coordination sphere and focus on the types of bonding involved in the coordination compound. The correct answer was selected by 62% of the students.

4. Determine the coordination number and oxidation number, respectively, of the transition metal in the coordination compound $[Cr(en)_2Br_2] NO_2$. The abbreviation "en" represents the neutral, bidentate ligand ethylenediamine.

A) 4, 3   B) 4, 2   C) 6, 3   D) 6, 2

Correct Answer: C
Level of Difficulty: Moderate/Challenging
Keyword(s): Coordination number and oxidation number
Histogram:

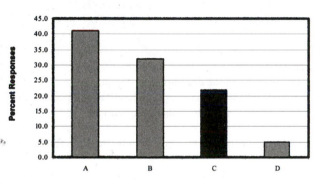

Comments: As the histogram indicates, only 22% responded correctly after a discussion of coordination number and oxidation number of the metal ion in the coordination compound. Some of the difficulty may arise from the concept of a bidentate ligand (41% selected distractor [A]) and 32% selected distractor (B). Students gain confidence with these "apparently" simple concepts as they work with more examples; mandatory homework is highly recommended and several excellent electronic homework systems are currently available.

5. Calculate the number of d electrons for cobalt in the complex ion: $[Co(NH_3)_2(en)Cl_2]^+$. The abbreviation "en" represents the neutral, bidentate ligand ethylenediamine.

A) 5   B) 6   C) 7   D) 8

Correct Answer: B
Level of Difficulty: Moderate
Comments: Most students recognized the neutral ligands and calculated the oxidation number of cobalt correctly, leading to the correct calculation of the number of d electrons.

6. What is the correct name of the compound $[Co(NH_3)_2(en)Cl_2]NO_3$?

        A) Diammoniadichloroethylenediaminecobalt(II) nitrate
        B) Diamminedichloroethylenediaminecobalt(III) nitrate
        C) Dichlorodiammineethylenediaminecobalt(II) nitrate
        D) Dichlorodiammineethylenediaminecobalt(III) nitrate

Correct Answer: B
Level of Difficulty: Easy/Moderate
Keyword(s): Naming coordination compounds
Comments: This is a relatively simple concept for most students and 84% responded correctly.

7. What is the correct formula of the compound potassium aquapentacyanocobaltate(III) ?

    A) $K[(H_2O)(CN)_5Co]$ B) $K[Co(H_2O)(CN)_5]$
    C) $K_2[Co(CN)_5(H_2O)]$ D) $K_2[Co(H_2O)(CN)_5]$

Correct Answer: D
Level of Difficulty: Easy
Keyword(s): Writing correct formulas for coordination compounds
Comments: This is also a relatively simple concept; 92% responded correctly. The convention places neutral ligands before anionic ligands in the complex ion.

8. How many geometric isomers of the square planar complex $[Pt(NH_3)(H_2O)BrCl]$ are possible?

    A) 1   B) 2   C) 3   D) 4

Correct Answer: C
Level of Difficulty: Moderate
Keyword(s): Geometric isomerism in square planar complexes
Histogram:

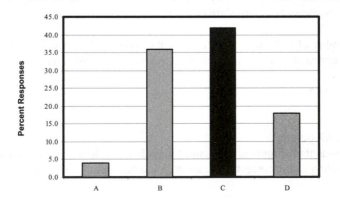

Comments: We were a little surprised by the results; 42% responded correctly and 36% identified only 2 of the 3 possible geometric isomers.

9. Predict the number of geometric isomers of the octahedral complex ion $[Co(en)_2Cl_2]^+$. Ethylene diamine is a *cis* ligand.

<blockquote>
A) 2   B) 3   C) 4   D) 5
</blockquote>

Correct Answer: A
Level of Difficulty: Easy/Moderate
Keyword(s): Geometric isomerism in octahedral complexes
Histogram:

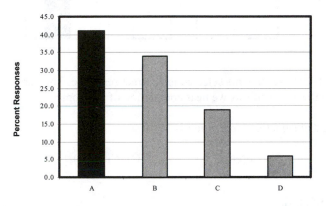

Comments: The histogram shows that the students struggled with this question even though they were informed that ethylene diamine is a *cis* ligand. Learning to visualize molecules in three dimensions is a challenge; only 41% selected the correct answer. Our students perform a molecular modeling lab exercise and have expressed their satisfaction with this learning activity.

10. Which complex species will exhibit optical isomerism?

<blockquote>
A) $[Co(en)Cl_4]^-$ B) *trans*-$[Cr(en)_2BrCl]^+$

C) *cis*-$[Co(NH_3)_4Cl_2]^+$ D) *cis*-$[C(ox)_2Br_2]^-$ (ox = oxalate, a bidentate ligand)
</blockquote>

Correct Answer: D
Level of Difficulty: Moderate/Challenging
Keyword(s): Optical isomerism in octahedral complexes

Note: Histogram and comments for this question appear on the next page.

Histogram:

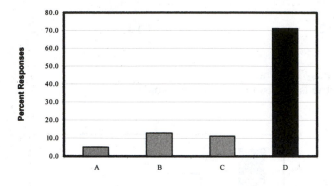

Comments: About 71% of the students selected the correct response. They appear to gain confidence with visualization skills as we near completion of this topic.

11. An example of a diamagnetic complex ion is

A) $[FeF_6]^{3-}$ B) $[Co(en)_3]^{2+}$ C) $[Cr(H_2O)_6]^{3+}$ D) $[Co(CN)_6]^{3-}$

Correct Answer: D
Level of Difficulty: Moderate
Keyword(s): Magnetic properties of transition metal complexes
Comments: About 73% responded correctly after a discussion of crystal-field theory and its application to the magnetic properties of transition metal complexes.

12. $[Ni(CN)_4]^{2-}$ is a diamagnetic complex while $[NiCl_4]^{2-}$ is a paramagnetic complex. Which statement best explains this experimental observation?

A) CN^- is a strong-field ligand while Cl^- is a weak-field ligand.
B) Cl^- is a strong-field ligand while CN^- is a weak-field ligand.
C) $[Ni(CN)_4]^{2-}$ is tetrahedral while $[NiCl_4]^{2-}$ is square planar.
D) $[Ni(CN)_4]^{2-}$ is square planar while $[NiCl_4]^{2-}$ is tetrahedral.

Correct Answer: D
Level of Difficulty: Challenging
Keyword(s): Square planar and tetrahedral complexes
Comments: This is a very challenging question; only 11% answered correctly and 65% selected distractor (A).

CHAPTER 26 Oxidation-Reduction Processes and Electrochemistry

1. What is the oxidation number of chromium in ammonium dichromate?

        A) +3   B) +4   C) +5   D) +6

Correct Answer: D
Level of Difficulty: Easy/Moderate
Keyword(s): Oxidation numbers
Histogram:

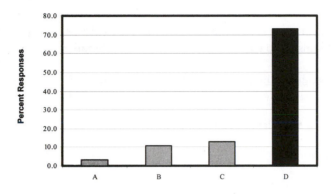

Comments: Only 73% responded correctly; some students had trouble writing down the correct formula for ammonium dichromate even though we provided information about the polyatomic ions.

2. Select all redox reactions by looking for a change in oxidation number as reactants are converted to products.

        I) $Ca + 2H_2O \rightarrow Ca(OH)_2 + H_2$
        II) $CaO + H_2O \rightarrow Ca(OH)_2$
        III) $Ca(OH)_2 + H_3PO_4 \rightarrow Ca_3(PO_4)_2 + H_2O$
        IV) $Cl_2 + 2KBr \rightarrow Br_2 + 2KCl$

        A) I and II   B) II and III   C) I and IV   D) III and IV

Correct Answer: C
Level of Difficulty: Moderate
Keyword(s): Using oxidation numbers to recognize redox reactions

Note: *Histogram and comments for this question appear on the next page.*

Histogram:

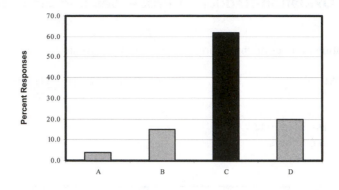

Comments: Some students may need more time to apply the oxidation number rules to several equations. About 62% answered correctly even with peer collaboration.

3. Dichromate ion in <u>acidic medium</u> converts ethanol, C_2H_5OH, to CO_2 according to the unbalanced equation: $Cr_2O_7^{2-}(aq) + C_2H_5OH(aq) \rightarrow Cr^{3+}(aq) + CO_2(g) + H_2O(l)$

The <u>coefficient for H^+</u> in the balanced equation using smallest integer coefficients is:

A) 8   B) 10   C) 13   D) 16

Correct Answer: D
Level of Difficulty: Moderate/Challenging
Keyword(s): Balancing redox equations in acidic medium
Histogram:

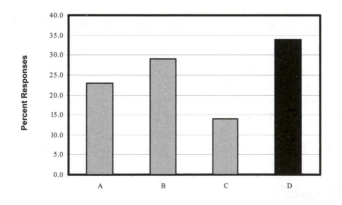

Comments: This was a challenging exercise for our students and only 34% answered correctly. Students need drill and practice with balancing redox equations.

4. Which species is the oxidizing agent in the reaction

$$Cr_2O_7{}^{2-}(aq) + C_2H_5OH(aq) \rightarrow Cr^{3+}(aq) + CO_2(g) + H_2O(l)?$$

A) $Cr^{3+}(aq)$ B) $Cr_2O_7{}^{2-}(aq)$ C) $CO_2(g)$ D) $C_2H_5OH(aq)$

Correct Answer: B
Level of Difficulty: Easy/Moderate
Keyword(s): Oxidizing and reducing agents
Histogram:

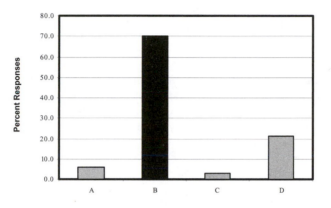

Comments: Students often feel more comfortable identifying oxidizing and reducing agents after they set up the half-reactions. We try to guide our students by reminding them that the oxidizing agent facilitates the oxidation process by removing electrons produced in the oxidation process; hence, the oxidizing agent is reduced. The correct answer was selected by 70% of the students, and 21% selected the other reactant.

5. The following reaction occurs in <u>basic medium</u>:

$$Zn(s) + NO_3{}^-(aq) \rightarrow Zn(OH)_4{}^{2-}(aq) + NH_3(aq)$$

The oxidizing agent is:

A) $Zn(s)$ B) $NO_3{}^-(aq)$ C) $OH^-(aq)$ D) $H_2O(l)$

Correct Answer: B
Level of Difficulty: Easy/Moderate
Keyword(s): Oxidizing and reducing agents

Note: Histogram and comments for this question appear on the next page.

Histogram:

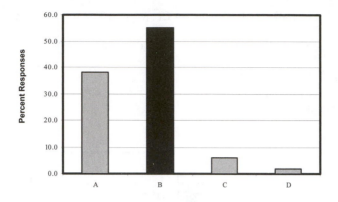

Comments: This question was presented at the start of the second lecture on oxidizing and reducing agents to review and reinforce concepts. About 55% responded correctly and 38% selected distractor (A) corresponding to the other reactant. Most students usually get a similar question correct on the exam confirming that they benefit from drill-and-practice on tutorials and electronic homework assignments.

6. Which statement about voltaic cells is not true?

    A) Reduction occurs at the cathode.
    B) Anions move in the salt bridge toward the electrode where oxidation is occurring.
    C) The electrode where reduction occurs is represented with a positive sign.
    D) Electrons flow in the external circuit from cathode to anode.

Correct Answer: D
Level of Difficulty: Easy
Keyword(s): Voltaic cells
Comments: Following the class discussion on voltaic cells, 94% responded correctly.

7. Metallic copper does not react with hydrochloric acid, but is oxidized by nitric acid since the nitrate ion functions as the oxidizing agent in acid medium. If NO and Cu^{2+} are formed as products, and the reaction is carried out under standard conditions, ε^0 for the reaction will be:

    A) +0.44 V   B) +0.62 V   C) +0.78 V   D) +0.96 V

Correct Answer: B
Level of Difficulty: Easy
Keyword(s): Standard reduction potentials and cell potential
Comments: The students need access to a table of standard reduction potentials. We have found that students are often confused by sign conventions. Many authors associate the symbol ε^0 for half-cell reactions with standard reduction potentials and prefer to represent the standard half-cell potential for the anode reaction (oxidation) as $-\varepsilon^0_{anode}$. This representation confuses some students,

especially when the textbook presents it differently from the instructor. We were quite pleased when 89% answered this question correctly.

8. Given: $Cu^{2+} + 2e^- \rightarrow Cu$, $\varepsilon^0 = +0.34$ V, and $NO_3^- + 4H^+ + 3e^- \rightarrow NO + 2H_2O$, $\varepsilon^0 = +0.96$ V, calculate the value of ε^0, in volts, for the reaction $3Cu + 2NO_3^- + 8 H^+ \rightarrow 3Cu^{2+} + 2NO + 4H_2O$.

A) −0.62 V   B) +0.62 V   C) −0.90 V   D) +0.90 V

Correct Answer: B
Level of Difficulty: Easy/Moderate
Keyword(s): Standard reduction potentials
Histogram:

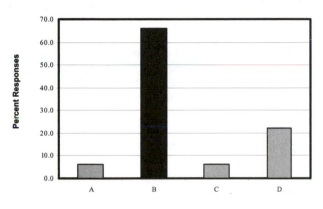

Comments: This question is almost identical to the previous question and was presented at the start of the second lecture on standard reduction potentials. Only 66% responded correctly compared to 89% when the concept was just presented in class. Performance on clicker tests reflects short-term understanding; long-term retention requires extensive drill and practice as well as review and reinforcement.

9. An electrochemical cell is constructed using the reaction:

$Zn(s) + I_2(s) \rightarrow Zn^{2+}(aq) + 2I^-(aq)$ Data: $\varepsilon^0(Zn^{2+}/Zn) = -0.76$ V; $\varepsilon^0(I_2/I^-) = +0.53$ V

Which statements about this cell are <u>correct</u>?

I) This is a voltaic cell.
II) A solid zinc electrode is the anode.
III) A solid iodine electrode is the cathode.
IV) Electrons will flow from the zinc electrode, through the wire, to the cathode.
V) Iodine is the reducing agent.

A) I, II & III   B) III, IV & V   C) I, II & IV   D) II, III & IV

Correct Answer: C

Level of Difficulty: Moderate/Challenging
Keyword(s): Voltaic cells
Histogram:

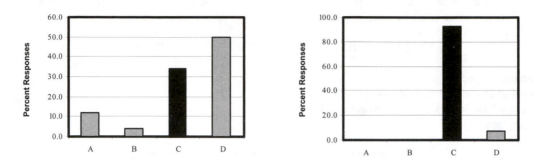

Comments: This question integrates many of the concepts about voltaic cells presented in class. The histogram on the left represents individual performance without peer collaboration; 34% selected the correct answer. After peer collaboration, 93% responded correctly as depicted in the histogram on the right. This exercise once again supports our practice of encouraging peer collaboration in class.

10. Examine the following half-reactions and identify the <u>strongest reducing agent</u>.

$$SnO_2(s) + 2H_2O(l) + 4e^- \rightarrow Sn(s) + 4OH^-(aq) \qquad \varepsilon^0 = -0.945 \text{ V}$$
$$Hg_2SO_4(s) + 2e^- \rightarrow 2Hg(l) + SO_4^{2-}(aq) \qquad \varepsilon^0 = +0.613 \text{ V}$$
$$Cr(OH)_3(s) + 3e^- \rightarrow Cr(s) + 3OH^-(aq) \qquad \varepsilon^0 = -1.48 \text{ V}$$
$$MnO_2(s) + 4H^+(aq) + 2e^- \rightarrow Mn^{2+}(aq) + 2H_2O(l) \qquad \varepsilon^0 = +1.224 \text{ V}$$

The strongest reducing agent is:

        A) Cr   B) MnO_2 C) Hg_2SO_4 D) Sn

Correct Answer: A
Level of Difficulty: Moderate/Challenging
Keyword(s): Relative oxidizing and reducing strengths

Note: Histogram and comments for this question appear on the next page.

Histogram:

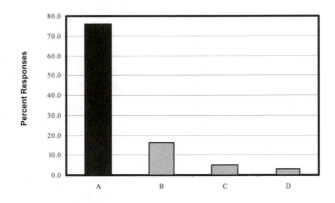

Comments: This question required students to recognize that reducing agents are oxidized; hence, they must consider the reverse reactions, keep track of the correct sign for ε^0, and then compare ε^0 values. The correct answer was selected by 76% of the students.

11. For a spontaneous electrochemical reaction under standard-state conditions,

A) $\Delta G^0 < 0$, $\varepsilon^0_{cell} < 0$, and $K < 1$
C) $\Delta G^0 > 0$, $\varepsilon^0_{cell} > 0$, and $K < 1$

B) $\Delta G^0 < 0$, $\varepsilon^0_{cell} > 0$, and $K > 1$
D) $\Delta G^0 > 0$, $\varepsilon^0_{cell} < 0$, and $K > 1$

Correct Answer: B
Level of Difficulty: Easy/Moderate
Keyword(s): Relationships between ΔG^0, ε^0_{cell}, and K
Comments: This question followed the discussion of the relationships between ΔG^0, ε^0_{cell}, and K, and 88% responded correctly. The students appear to have been engaged in the lecture.

12. Predict the value of ε^0_{cell} for the cell: Cu(s)/Cu^{2+}(0.10 M)// Cu^{2+}(1.0 M)/ Cu(s).

Given: $Cu^{2+} + 2e^- \rightarrow Cu$, $\varepsilon^0 = +0.34$ V

A) $\varepsilon^0_{cell} = +0.34$ V
C) $\varepsilon^0_{cell} = -0.34$ V

B) $\varepsilon^0_{cell} = 0$
D) More information is required

Correct Answer: B
Level of Difficulty: Moderate
Keyword(s): Concentration cells

Note: Histogram and comments for this question appear on the next page.

Histogram:

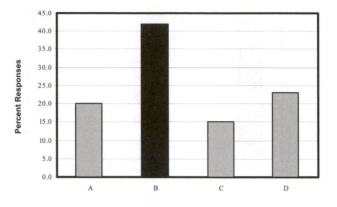

Comments: This question was written just prior to walking into the lecture hall. We sometimes assume that some concepts are quite obvious; often this is a serious mistake and an eye-opener for the instructor. We remembered that students frequently forget that ε^0_{cell} is zero for a concentration cell. The question was presented after an introduction to concentration cells using a visual, emphasizing that the half-cells are identical in all respects except for the concentration. The histogram shows that only 42% applied the concepts correctly; 35% thought that the value was either +0.34 V or −0.34 V.

13. For a concentration cell: $\varepsilon_{cell} = \dfrac{-0.0592 \text{ V}}{n} \log Q$

At 25°C, a cell has $[Ag^+] = 1.00$ M in half-cell **A** and $[Ag^+] = 0.100$ M in half-cell **B**. Which answer includes <u>ALL correct</u> statements?

> I) The reaction continues until $[Ag^+] = 0.7$ M in each half-cell.
> II) Half-cell **A** contains the anode.
> III) The cell potential is 0.0592 V.
> IV) $\varepsilon^0_{cell} = 0$

> A) I and II   B) III and IV   C) I, III and IV   D) II, III and IV

Correct Answer: B
Level of Difficulty: Moderate
Keyword(s): Concentration cells and the Nernst equation
Comments: This question tested several concepts related to concentration cells and it was encouraging to find that 90% responded correctly. The use of clickers is highly recommended as it provides information in real time about the teaching and learning processes.

14. Voltaic cells and electrolytic cells are based on thermodynamic principles. Which statement about these cells is <u>correct</u>?

    A) The mass of the anode increases in a voltaic cell as the cell discharges.
    B) Reduction occurs at the anode only in the electrolytic cell.
    C) In a voltaic cell, electrons travel from the cathode to the anode in solution.
    D) The free energy change ΔG is greater than zero for the electrolytic cell.

Correct Answer: D
Level of Difficulty: Moderate
Keyword(s): Comparison of voltaic and electrolytic cells
Comments: About 15% of the students were tripped by the statement "Reduction occurs at the anode only in the electrolytic cell"; 80% responded correctly.

15. How many grams of Cr would plate out from a solution of $Cr(NO_3)_3$ when 1.93×10^5 coulombs of charge are passed through the solution? The atomic mass of Cr is 52.0 g/mol, and 1 Faraday is equal to 9.65×10^4 C/mol e^-.

    A) 17.3 g   B) 34.7 g   C) 52.0 g   D) 104 g

Correct Answer: B
Level of Difficulty: Moderate
Keyword(s): Faraday's laws of electrolysis
Comments: Some students have a difficult time writing the correct equation for the half-reaction and integrating the qualitative and quantitative concepts. About 81% selected the correct answer.

CHAPTER 27 Nuclear Chemistry

1. Isotopes of an element differ in the number of

  A) electrons.   B) neutrons.
  C) protons.   D) electrons, neutrons, and protons.

Correct Answer: B
Level of Difficulty: Easy
Keyword(s): Isotopes
Histogram:

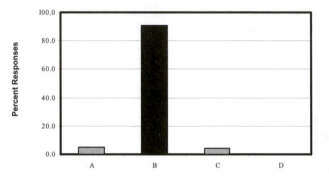

Comments: This question tests prior knowledge and 91% responded correctly.

2. How many neutrons and protons are there in the radioisotope ^{60}Co that is used in cancer therapy? The atomic number of cobalt is 27.

  A) 60 neutrons and 27 protons   B) 27 neutrons and 60 protons
  C) 33 neutrons and 27 protons   D) 27 neutrons and 33 protons

Correct Answer: C
Level of Difficulty: Easy
Keyword(s): Isotope symbols

Note: Histogram and comments for this question appear on the next page.

178

Histogram:

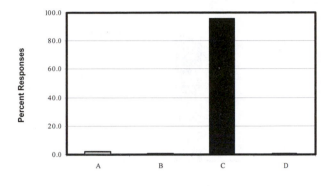

Comments: This question also tests prior knowledge and 96% responded correctly, confirming our experience that students master these skills. Review and reinforcement establish a solid foundation that assists in designing lectures on nuclear chemistry.

3. Select the <u>incorrect</u> statement about radioactive emissions.

        A) In positron emission, a neutron within the nucleus is converted to a proton.
        B) Order of penetrating power of emissions is $\gamma > \beta > \alpha$.
        C) β-decay results in a product nuclide with the same mass number but
           an atomic number one higher than the reactant nuclide.
        D) X-ray emission accompanies electron capture.

Correct Answer: A
Level of Difficulty: Moderate
Keyword(s): Radioactive decay and properties of particles and rays
Comments: This question captures some basic ideas about radioactive decay.

4. What is the missing nuclide X in the nuclear reaction $^{96}Mo + {}^{2}H \rightarrow X + $ neutron? The atomic numbers for niobium, molybdenum, and technetium are 41, 42, and 43, respectively.

        A) ^{97}Mo B) ^{97}Tc C) ^{97}Nb D) ^{98}Tc

Correct Answer: B
Level of Difficulty: Moderate
Keyword(s): Nuclear equations

Note: Histogram and comments for this question appear on the next page.

Histogram:

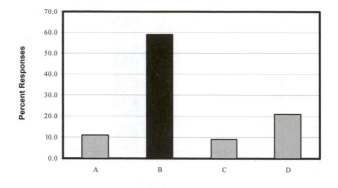

Comments: The students are initially challenged by the logic of balancing equations for nuclear reactions. Only 59% selected the correct answer. The drill and practice provided by electronic homework prepares our students to be successful on similar types of questions on exams.

5. Polonium-210 (atomic number 84) decays by α-particle emission. What is the product of this nuclear decay process? The atomic numbers of mercury, lead, radon, and radium are 80, 82, 86, and 88, respectively.

A) ^{206}Pb   B) ^{206}Rn   C) ^{208}Ra   D) ^{208}Hg

Correct Answer: A
Level of Difficulty: Easy/Moderate
Keyword(s): Writing nuclear equations
Comments: This question provides more practice on writing the nuclear equation when α-particle emission occurs. It will be easy for the students who have mastered the logic and another learning opportunity for those who need more practice.

6. Product X is formed from ^{233}Np (atomic number is 93) by electron capture. What is the chemical identity of X?

A) $_{91}$Pa   B) $_{92}$U   C) $_{93}$Np   D) $_{94}$Pu

Correct Answer: B
Level of Difficulty: Easy
Keyword(s): Nuclear Equations

Note: Histogram and comments for this question appear on the next page.

Histogram:

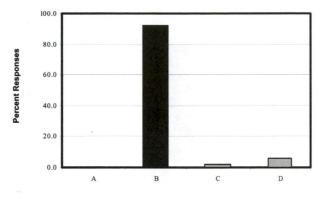

Comments: This histogram shows that 92% of the students show some understanding of the concepts involved in writing a correct nuclear equation. It also provides a variation in the types of questions that can be presented on homework, quizzes, and exams.

7. X is a positron emitter. Which of these isotopes might X be?

    A) ^{18}N   B) ^{23}F   C) ^{15}O   D) ^{36}P

Correct Answer: C
Level of Difficulty: Moderate
Keyword(s): Nuclear stability, belt of stability and predicting decay mode
Comments: This topic has been challenging for our students. It is important to provide access to the periodic table so that students can use atomic numbers and the information about average atomic mass as a guideline. In many cases, the average atomic mass is closer to that of the more stable isotope. In fluorine, the weighted average atomic mass is 19; hence, the isotope ^{18}F has a lower neutron-to-proton ratio and the net change is the conversion of a proton to a neutron and the emission of a positron.

8. Predict the most likely mode of decay for ^{15}C.

              A) α-particle emission   B) β-particle emission
              C) neutron capture   D) positron emission or electron capture

Correct Answer: B
Level of Difficulty: Easy/Moderate
Keyword(s): Belt of stability and predicting decay mode

Note: Histogram and comments for this question appear on the next page.

Histogram:

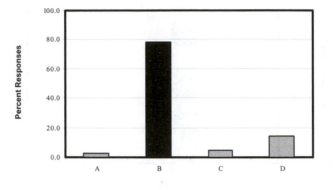

Comments: This question was answered correctly by 78% of the students showing that they are gaining more confidence in predicting mode of decay of a radioactive isotope. The burning question is always whether the figure of the belt of stability will be provided on the exam. We train our students to extract information from the periodic table.

9. Cesium-133 is a stable nuclide, but Cesium-123 is radioactive. What is the most likely mode of decay of ^{123}Cs? Atomic number is 55.

A) α-decay

B) β-decay

C) Positron emission/electron capture

D) Proton emission

Correct Answer: C
Level of Difficulty: Easy/Moderate
Keyword(s): Predicting mode of decay
Histogram:

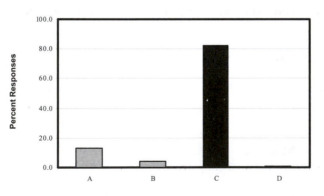

Comments: The mass number of the stable isotope was provided in the question and 82% answered correctly.

10. Which radioactive decay prediction is very likely not correct?

A) ^{47}Ca, beta decay   B) ^{25}Al, positron emission
C) ^{230}Th, alpha decay   D) ^{24}Ne, electron capture

Correct Answer: D
Level of Difficulty: Moderate/Challenging
Keyword(s): Predicting decay mode using belt of stability
Histogram:

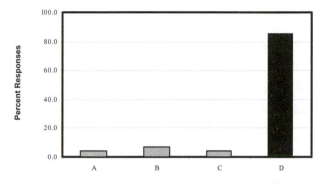

Comments: This question summarizes the information discussed about the various modes of decay and prediction of the decay mode using ideas contained in the nuclear belt of stability. About 85% selected the correct answer. Although this topic starts out as a challenge, many students feel confident after the drill and practice in class using clicker questions.

11. Gallium-65, a radioactive isotope of gallium, decays by first-order kinetics. The half-life of this isotope is 15.2 minutes. How long would it take for ⅞ of a sample of this isotope to decay?

A) 15.2 min   B) 30.4 min   C) 45.6 min   D) 48.0 min

Correct Answer: C
Level of Difficulty: Easy/Moderate
Keyword(s): Half-life
Comments: This question could be relatively easy if students grasped the meaning of half-life and recognized that one-eighth of the sample remains after 3 half lives. A similar question was given on an hour exam several years ago, and less than half of the class selected the correct answer. Perhaps the students do not clearly distinguish between the amount that decays and the amount that remains at a certain point in time.

CHAPTER 28 Metals and Nonmetals

1. Which element forms an amphoteric oxide?

   A) Al   B) Ca   C) K   D) S

Correct Answer: A
Level of Difficulty: Moderate/ Challenging
Keyword(s): Oxides of metals and nonmetals
Comments: This question tests information learned in first-semester general chemistry about periodicity of properties. Our experience shows that students tend not to remember these details of descriptive inorganic chemistry even though we attempt to reinforce this information throughout the two semesters.

2. Which metal is most likely to occur in nature in the elemental form?

   A) Au   B) Ca   C) Na   D) Zn

Correct Answer: A
Level of Difficulty: Easy/Moderate
Keyword(s): Occurrence of metals in nature
Comments: Students should recognize that Au is the symbol for gold and associate inertness with this metal, accounting for its natural occurrence in the elemental state.

3. Which metal does not dissolve in aqueous HCl?

   A) Al   B) Cu   C) Mg   D) Zn

Correct Answer: B
Level of Difficulty: Easy/Moderate
Keyword(s): Reactions of metals with acids
Comments: This question provides an opportunity to make connections to the chapter on redox reactions and electrochemistry. Many students perform an experiment where they react metals with HCl and observe fast and slow reactions, accompanied by the evolution of H_2 gas. Copper reacts only with oxidizing acids such as HNO_3.

4. Which statement about the transition elements is <u>incorrect</u>?

  A) Transition elements form more coordination compounds than the elements
   in groups IA and IIA.
  B) All coordination compounds containing transition elements are paramagnetic.
  C) Many coordination compounds containing transition elements have absorption
   bands in the visible region of the spectrum.
  D) All transition elements are metals.

Correct Answer: B
Level of Difficulty: Moderate
Keyword(s): Transition elements
Comments: This question reviews some of the information presented in the chapter on transition metals.

5. Which of these is <u>not</u> an allotrope of carbon?

   A) Graphite   B) Diamond   C) Benzene   D) Buckminsterfullerene

Correct Answer: C
Level of Difficulty: Easy/Moderate
Keyword(s): Allotropes, allotropes of carbon
Comments: This question may be used to test prior knowledge before a discussion of the allotropes of carbon and the applications of these.

6. Which statement is <u>incorrect</u>?

   A) Diborane, B_2H_6, is described as a molecule with three-center, two-electron bridge bonds.
   B) Boron is a metalloid.
   C) Silicon-silicon multiple bonds are rare compared to carbon-carbon multiple bonds.
   D) CO_2 and SiO_2 have similar bonding and structure.

Correct Answer: D
Level of Difficulty: Moderate/Challenging
Keyword(s): Boron, carbon, silicon, and their compounds
Comments: This question may be used to summarize the chemistry of B, C, Si and their compounds.

7. Which statement is <u>correct</u>?

   A) The earth's atmosphere transmits infrared light and traps visible light.
   B) The concentration of CO_2 in the atmosphere undergoes seasonal oscillations.
   C) Deforestation will counteract global warming.
   D) The major greenhouse gases are N_2 and O_2.

Correct Answer: B
Level of Difficulty: Moderate
Keyword(s): CO_2 and the greenhouse effect
Comments: We often present one lecture in the second semester on the greenhouse effect. This question could follow such a discussion.

8. Which carbon-containing species contains only sp^3-hybridized carbon atoms?

   A) CO   B) CO_2 C) diamond   D) graphite

Correct Answer: C
Level of Difficulty: Easy/Moderate
Keyword(s): Carbon chemistry
Comments: Some basic concepts about chemical bonding can be tested in the context of the allotropes and oxides of carbon.

9. Which nitrogen-containing species is least reactive?

A) N_2 B) NO   C) NO_2 D) NH_4NO_3

Correct Answer: A
Level of Difficulty: Easy/Moderate
Keyword(s): Nitrogen and its compounds
Comments: This is an excellent question to reinforce the stability of the triple bond in the N_2 molecule and the odd-electron nature of NO and NO_2. We would also discuss the explosive nature of some nitrogen compounds such as NH_4NO_3 and nitroglycerin.

10. The element berkelium, first prepared in 1949 at the University of California at Berkeley, is made by the reaction ^{241}Am $(\alpha, 2n)$ ^{243}Bk. This is an example of

A) nuclear fusion.   B) nuclear fission.
C) nuclear transmutation.   D) stellar nucleogenesis.

Correct Answer: C
Level of Difficulty: Moderate
Keyword(s): Nuclear fusion, fission, transmutation, and stellar nucleogenesis
Comments: This question could be used in the chapter on nuclear chemistry. Students need access to the periodic table to answer this question.

11. Which of these is a transuranium element?

A) Ra   B) Rn   C) Tc   D) Cm

Correct Answer: D
Level of Difficulty: Easy
Keyword(s): Transuranium elements
Comments: This question could also be used in the chapter on nuclear chemistry. Students need access to the periodic table.

CHAPTER 29 Lecture Demo Instructions and Preparation Notes

I have successfully used the following demonstrations with clicker questions in my introductory general chemistry lectures. So that you may use them in yours, I have provided setup and demonstration instructions, safety tips, and suggestions on how to integrate a particular clicker question with each demonstration. You can find many more lecture demonstrations at http://www.colorado.edu/chem/genchem/demoman/start.html.

Physical Change: Sublimation of Iodine
Iodine crystals are heated (directly or indirectly) and the vapors are condensed to the solid. Both the vapor and crystals formed by deposition may be observed.

Use with Clicker Question: Chapter 7, Question 2
I would start the demo and inform my students that we were observing a physical change using the sublimation of iodine as an example. It would be ideal to project the demo using a camera in a large classroom. I would then project the clicker question and encourage students to make connections between the macroscopic observations and the symbolic representations in equations. You can design a related question that addresses the molecular view of the sublimation process. This demo and the conceptual question(s) allow you to review phase transitions in the context of vocabulary such as sublimation and deposition.

Setup Prior to Lecture
A. Assemble the following:
    a.   Iodine crystals in sealed glass tube
    b.   Hot plate
    c.   Tall 1000-mL beaker half full of water

OR

B. Assemble the following:
    a.   Iodine crystals
    b.   600-mL Beaker
    c.   500-mL Florence flask
    d.   Crushed ice
    e.   Hot plate

Demonstration Steps
A. 1. Heat a tall 1000-mL beaker full of water to boiling.
     2. Place a closed glass tube of iodine crystals in the water. The tube will fill up with iodine vapor. No liquid iodine will form.

OR

B.   1. Fill a Florence flask with ice and place over a beaker containing a small amount of iodine crystals.
     2. Heat on *low heat* over a hot plate. Iodine will sublime and form crystals on the cooled bottom (outside) of the Florence flask.

Safety and Disposal
Do not inhale iodine vapors.

Physical Change: Sublimation of Dry Ice
Small pieces of dry ice are carefully placed in a disposable latex glove and the opening is tied. As dry ice sublimes, the glove expands.

Use with Clicker Question: Chapter 7, Question 3
I would start the demo and inform my students that we were observing a physical change using the sublimation of iodine as an example. I would then project the clicker question and encourage students to make connections between the macroscopic observations and the symbolic representations in equations. You can design a related question that addresses the molecular view of the sublimation process. This demo and the conceptual question(s) allow you to review phase transitions in the context of vocabulary such as sublimation and deposition.

Setup Prior to Lecture
Assemble the following:
    a.   Disposable latex gloves
    b.   Small (golf ball or smaller) pieces of dry ice
    c.   Protective gloves

Demonstration Steps
1. Place a chunk of dry ice into a disposable latex glove. Tie the opening so that no gas escapes.
2. After the glove fills with CO_2, pass it around the room to show that it contains no liquid.

Safety and Disposal
Dry ice can cause burns to exposed skin. Wear protective gloves when handling dry ice. Do not breathe iodine vapors. Dispose of iodine following proper disposal practices at your school or university. Do not flush down the sink.

Chemical Change: Electrolysis of Water
Electrical current is applied to water containing Na_2SO_4 as an electrolyte in a specially designed electrolysis apparatus for use on an overhead projector. The water is converted into gaseous H_2 and O_2 at the platinum electrodes.

Use with Clicker Question: Chapter 7, Question 5
I would start the demo and draw attention to the color of the indicator in the solution and the two electrodes in the apparatus. I would encourage students to observe the region surrounding each electrode as the electrolysis proceeds. I would then project the clicker question and ask students to

make connections between the macroscopic observations and the symbolic representations in equations. You can design a related question that addresses the molecular view of electrolysis.

NOTE: This demo may also be used in the electrochemistry unit to link the observations at each electrode with the respective half-reaction (oxidation or reduction).

Setup Prior to Lecture
1. Prepare 100 mL of 1M sodium sulfate (Na_2SO_4) solution and add several drops of universal indicator. (Note: add 1 drop of 0.1 M NaOH to solution if color is pink—the color should be green to start demo.)

$$14.2 \text{ g } Na_2SO_4 / 100 \text{mL}$$

2. Wire two 9-volt batteries together in series. Fix alligator clips on the ends of the wires.
3. Assemble the following:
   a. Electrolysis apparatus filled with electrolyte
   b. Petri dish
   c. Paper towels

Demonstration Steps
1. Attach one alligator clip from the battery pack to each electrode.
2. Oxygen and H^+ are formed at the anode (oxidation), and the indicator turns red. Hydrogen and OH^- are formed at the cathode (reduction), and the indicator turns blue. After a few minutes, a relative volume ratio of 2:1 for H_2 to O_2 should be visible.

Equations
Anode Half Reaction: $2 H_2O(l) \rightarrow 4 H^+(aq) + O_2(g) + 4 e^-$
Cathode Half Reaction: $4 H_2O(l) + 4 e^- \rightarrow 2 H_2(g) + 4 OH^-(aq)$
Overall Reaction: $2 H_2O(l) \rightarrow 2 H_2(g) + O_2(g)$

Safety and Disposal
The electrolysis apparatus is somewhat fragile, and care should be taken in packing and transporting it.

Acid-Base and Precipitation Reaction: Using Conductivity
A conductivity apparatus is used to monitor the neutralization of barium hydroxide with sulfuric acid.

Use with Clicker Question: Chapter 11, Question 6
After a discussion on strong and weak electrolytes and non-electrolytes, I would project this question and have students predict the outcome of the reaction. This is a special case of an acid-base reaction that is accompanied by precipitation. After displaying the results histogram, I would discuss the observations in light of the presence or absence of ions in solution and which ions were involved. I would also review the solubility rules to explain the precipitation of $BaSO_4$.

Setup Prior to Lecture
1. Prepare the following chemicals:
   a. 2.0 g Ba(OH)$_2$·8H$_2$O in a 250-mL beaker (record weight and include)
   b. Approximately 50 mL of 0.25 M H$_2$SO$_4$
   c. Distilled water in wash bottle

2. Assemble the following equipment:
   a. 50-mL Buret
   b. Single-bulb conductivity apparatus
   c. Stir plate and stir bar
   d. Ring stand with buret clamp
   e. Goggles
   f. Gloves

Demonstration Steps
1. Place the single-bulb conductivity apparatus on top of a 250-mL beaker containing 2 g of barium hydroxide in 50 mL of distilled water. Stir with a magnetic stir bar and stir plate.
2. Position the buret containing 0.25 M sulfuric acid so that acid may be delivered through a hole in the base and into the beaker.
3. Plug in the conductivity apparatus. The light bulb should go on due to the conductivity of ionic barium hydroxide in solution. Begin adding sulfuric acid from the buret. Allow several seconds after each addition.
4. As the titration progresses, insoluble barium sulfate will precipitate. The solution will no longer conduct electricity and the light bulb will go out and stay off in a range of approximately 1mL to either side of the endpoint. The endpoint should take approximately 25 mL of 0.25 M H$_2$SO$_4$ for a 2 g sample of Ba(OH)$_2$·8H$_2$O.
5. Add excess sulfuric acid to show that the light will go back on.

Equations
H$_2$SO$_4$(aq) + Ba(OH)$_2$(aq) → BaSO$_4$(s) + 2 H$_2$O(l)

Safety and Disposal
Care must be taken with the conductivity apparatus to prevent electrical shock. Barium salts are hazardous.

Stoichiometry: Limiting Reagent
Magnesium metal is dissolved in HCl in 500-mL Florence flasks covered with balloons. The hydrogen gas evolved is collected in the balloons, and the size of each balloon is proportional to the amount of hydrogen produced.

Use with Clicker Question: Questions Q4.4–7, Chapter 4 (pg. 19)
After a discussion of the limiting reactant concept, I would perform the demo and have students predict the outcomes. The clicker questions provide drill and practice with the limiting reactant concept. After each results histogram is displayed, I present the calculations to support the selection of the limiting reactant. On the last question, we calculate the theoretical amount of H$_2$ gas in each flask. At the end of the lecture, I go back to the demo and we check relative balloon

sizes for amounts of H_2 gas in each flask.

Setup Prior to Lecture

1. Prepare 800 mL of 0.50 M HCl (add 33 mL of 12 M HCl to 750 mL water and dilute to 800 mL) and divide evenly between four 500-mL Florence flasks (200 mL/0.10 mole each).
2. Weigh out the following amounts of magnesium turnings:

Flask 1	0.30g	0.0125 mole
Flask 2	0.61g	0.0250 mole
Flask 3	1.22g	0.0500 mole
Flask 4	2.43g	0.1000 mole

3. Pre-stretch the balloons. Blow them up to an equal size and let the air out.
4. Place correct amounts of Mg turnings in each of four balloons and carefully (Mg turnings must not drop into flask) stretch balloons over tops of Florence flasks containing HCl. Magnesium turnings should be added carefully to keep from puncturing balloons.

Demonstration Steps

1. Lift the balloons one at a time so that the Mg turnings fall into the HCl in each flask. Swirl to speed up reaction.
2. Each flask contains 0.1 mole of HCl. Magnesium is present in the following amounts:
   Flask 1: 0.0125 mole (0.30 g); excess HCl
   Flask 2: 0.0250 mole (0.61g); excess HCl
   Flask 3: 0.0500 mole (1.22 g); stoichiometric amount of HCl
   Flask 4: 0.1000 mole (2.43 g); excess Mg
3. Flasks 1 and 2 are limited by smaller stoichiometric amounts of Mg. Flask 3 will react to use both reagents completely. Flask 4 will produce only the same amount of hydrogen as Flask 3 and have excess Mg left over. HCl is the limiting reactant.

Equations

$Mg(s) + 2\ HCl(aq) \longrightarrow MgCl_2(aq) + H_2(g)$

Safety and Disposal

Handle acids with caution. Combine chemicals from all flasks to react away leftover magnesium.

Gases: Boyle's Law

An apparatus containing a clear glass syringe connected to a pressure gauge mounted on a heavy polycarbonate sheet is placed on the overhead projector to examine the relationship between pressure and volume of gas trapped in the syringe.

Use with Clicker Question: Chapter 13, Question 2

I perform this lecture demo to examine the relationship between pressure and volume of an air sample. Next, I project the clicker question. After displaying the results histogram, I remind the students that the amount of air and temperature were constant. I also review the concept of direct and inverse proportions. I draw graphs to represent the PV relationships.

Setup Prior to Lecture
1. Provide Boyle's Law apparatus.

Demonstration Steps
1. Take a series of measurements at different volumes. Record the volume and the resultant pressure. Observe relationship (inverse proportionality) between P and V for constant amount of gas at constant T.
2. Calculate individual products or plot V vs. 1/P to show the constant relationship PV = k.

Equations
PV = constant (k)   A plot of V vs. 1/P gives a straight line with slope of k.

Atomic Spectroscopy: Flame Tests
Salt slurries with methanol in Petri dishes are ignited and characteristic colors for various metals are observed.

Use with Clicker Question: Chapter 14, Question 4
After a discussion of atomic spectra, I perform the demo and follow it up with the clicker question. After displaying the results histogram, I remind students about the inverse relationship between energy and frequency. We also review the relative frequencies of the colors in the visible region of the electromagnetic spectrum.

Setup Prior to Lecture
1. Add 1-2 g of each salt to be displayed in a Petri dish with lid. Label both dish and lid with marker. Suggested metals are: Li (red), Ba (yellow), Ca (red and blue), Cu (green), K (lavender), Sr (orange-red), Na (orange), and Ni (blue and sparking). Allow salts to dry before storing Petri dishes.
2. Assemble the following:
    a.   Distilled water in wash bottle
    b.   Methanol in wash bottle
    c.   Matches
    d.   Fire extinguisher
    e.   Goggles

Demonstration Steps
1. Add a few milliliters of distilled water to each metal salt in its Petri dish and mix to form a slurry.
2. Pour methanol into each dish to a depth of $1/2$ inch.
3. Carefully ignite the alcohol with a match. After 15-60 seconds, colors should appear. If the flame burns out, add additional alcohol and re-ignite.

Additional Information
Colors of emission spectra are due to loosely bound valence electrons being excited in a flame and emitting light when dropping back to lower energy levels.

Safety and Disposal
Salts may spatter and eye protection should be worn. Place dishes on heatproof surface. Dishes remain hot after the flame goes out and should be handled with care.

Enthalpy, Entropy and Free Energy: Stretching Rubber Bands
Students stretch pre-cut rubber bands to their lips, stretch the rubber bands, and then allow them to contract. The rubber bands grow warmer as they expand and cool as they contract.

Use with Clicker Question: Chapter 20, Question 12
The clicker question is presented and students perform the demo to answer the question. Students enjoy participating in this exercise.

Setup Prior to Lecture
1. Cut enough #64 rubber bands (0.25" wide) into two straight-length strips to give one strip to each student.

Demonstration Steps
1. Pass out one rubber band to each student.
2. Instruct students to place rubber bands against their lips and stretch the bands, observing any temperature change.
3. Instruct students to allow rubber bands to contract, observing any temperature change.

Safety and Disposal
None.

Kinetics: CoCl$_2$-Catalyzed Oxidation of Tartaric Acid by H$_2$O$_2$
Upon addition of hydrogen peroxide to Rochelle's salt (sodium potassium tartrate), the hydrogen peroxide decomposes and tartaric acid is oxidized. This reaction proceeds very slowly at room temperature. The addition of cobalt chloride as a catalyst produces a green cobalt tartrate intermediate and copious evolution of oxygen and carbon dioxide. The pink color of the regenerated catalyst is visible upon completion of the reaction.

Use with Clicker Question: Chapter 21, Question 9
The clicker question on catalysts is first presented to test prior knowledge, as many students are familiar with the idea of a catalyst. The lecture demo is then performed to show that the catalyst speeds up the reaction and actually participates and is recovered upon completion of the reaction. This is a very effective lecture demonstration.

Setup Prior to Lecture
1. Assemble the following materials:
   a. 190 × 100 mm Crystallizing dish
   b. 600-mL Beaker for reaction
   c. 100-mL Beaker to display color of catalyst
   d. Stirring rod
   e. Hot plate

    f.   Thermometer
    g. Paper towels
    h. Gloves
    i.   Goggles

2. Prepare the following chemicals:
    a) 30 g Potassium sodium tartrate
    b) 30 mL of 30% hydrogen peroxide
    c) 50 mL of 0.4 M $CoCl_2$ (95.2 g $CoCl_2 \cdot H_2O$/L)
3. At the start of lecture, heat 400 mL of distilled water to 70°C in a 600-mL beaker. Dissolve potassium sodium tartrate in the water and maintain temperature until demonstration is performed.

Demonstration Steps
1. Place the 600-mL beaker containing 400 mL of sodium potassium tartrate solution at 70°C into a large crystallizing dish on the overhead projector. The temperature must be very close to 70°C for the reaction to proceed at an appropriate rate.
2. Add 30 mL of 30% hydrogen peroxide with stirring. Point out that the reaction is proceeding very slowly if at all (no gas bubbles visible).
3. Pour 50 mL of 0.4 M cobalt chloride solution into a beaker to demonstrate the color of the catalyst solution, and then add it to the reaction mixture with stirring.
4. Stand back! The reaction will be vigorous and solution will overflow the beaker into the crystallizing dish if the temperature is high enough. The green cobalt tartrate activated complex will be visible for the duration of the reaction.
5. Upon completion of the reaction, the solution will return to the pink color of the cobalt chloride catalyst, demonstrating that the catalyst was not consumed in the reaction.

Equation
$HO_2CCH(OH)CH(OH)CO_2H + H_2O_2 \rightarrow HO_2CCO_2H + O_2 + CO_2$
       tartaric acid   oxalic acid
NOTE: This reaction is not balanced. Other products are possible.

Safety and Disposal
Handle 30% hydrogen peroxide solution with care. Gloves are highly recommended. Watch out for spattering during the catalyzed reaction.

Chemical Equilibrium: Effect of Temperature on System 2 $NO_2 \rightleftharpoons N_2O_4$
Sealed tubes of a pale brown mixture of NO_2 and N_2O_4 (about equal intensity in tubes) are placed in hot and cold baths to shift the equilibrium. The coloration of the gas inside shows the effects. A third tube at room temperature is recommended for comparison. A fourth tube may be cooled in liquid N_2. This mixture appears colorless due to the excess of N_2O_4 and visually undetectable amounts of NO_2.

Use with Clicker Question: Chapter 22, Question 1
The demo is performed after the clicker question. We carry out the experiment at different temperatures and discuss the observations in the context of Le Châtelier's Principle because our

students perform the experiment on chemical equilibrium and Le Châtelier's Principle prior to the discussion in lecture. You can repeat this demo when you discuss Le Châtelier's Principle and the effect of temperature on chemical equilibria.

Setup Prior to Lecture
1. Assemble the following:
   a. Two or three sealed tubes of NO_2/N_2O_4
   b. Two tall 1000-mL beakers, one with hot water and one with ice water
OR
   a. One tall 1000-mL beaker with hot water and a tray of dry ice
   b. Hot plate
   c. Heat-protective gloves

Demonstration Steps
1. Place one tube in the cold bath. The equilibrium will shift toward N_2O_4 and the tube will lose color intensity.
2. Place one tube in the hot bath. The equilibrium will shift toward NO_2 and the color intensity will increase.

Equation
$$2\,NO_2(g) \rightleftharpoons N_2O_4(g) + D$$
  brown   colorless

Safety and Disposal
NO_2 is a toxic gas. In case the tubes break, do not inhale escaping gas. Wear heat-protective gloves when handling very cold or very hot tubes.

Complex Ion Equilibrium: Metal Ion-Ammonia Complex
Ammonia is added to a Cu^{2+} solution. Careful drop-wise addition produces a pale blue precipitate that dissolves in excess NH_3 to form a deep blue solution containing the complex ion.

Use with Clicker Question: Chapter 24, Question 9
The lecture demo is performed at the end of the clicker question to reinforce the pale blue precipitate of copper(II) hydroxide that dissolves in excess ammonia to form the deep blue $Cu(NH_3)_4^{2+}$ complex.

Setup Prior to Lecture
1. Prepare 0.1 M solutions of the following:
    $Cu(NO_3)_2\cdot2.5H_2O$ (copper nitrate)   MW = 232.6 g/mol
2. Assemble the following:
   a. 3 M or 6 M NH_4OH in a dropper bottle
   b. 15 M NH_4OH in a dropper bottle
   c. Two 50-mL beakers or 200-mm test tubes in lighted rack
   d. A stir rod
   e. Goggles
   f. Gloves

Demonstration Steps
1. Pour 0.1 M Cu^{2+} solution into a pair of labeled beakers (to a depth of 1/2") or into 200-mm test tubes (to a depth of 1") in a lighted rack.
2. Add 3 M or 6 M NH_4OH drop-wise while stirring with the stir rod to *one* beaker or test tube. The solution will become cloudy from the precipitation of the metal hydroxide.
3. Add 15 M NH_4OH with stirring to the same beaker or test tube. The precipitate will dissolve and a deep blue complex will form. Use the other beaker or test tube as reference.

Equations
$Cu^{2+}(aq) + NH_4OH(aq) \rightleftharpoons Cu(OH)_2(s)$; pale blue precipitate

$Cu(OH)_2(s) + 4\ NH_3(aq) \rightleftharpoons Cu(NH_3)_4^{2+}(aq)$; deep blue complex

Safety and Disposal
Avoid breathing concentrated ammonia vapors.

WORKS CITED

Asirvatham, M. R. (2005). "IR Clickers and ConcepTests: Engaging Students in the Classroom," CONFCHEM 2005 Winter Conference: "Trends and Innovations in Chemical Education" http://www.files.chem.vt.edu/confchem/2005/a/asirvatham.pdf, (accessed July 8, 2009)

Beatty, I. D. (2004). "Transforming Student Learning with Classroom Communication Systems," Research Bulletin ERB0403, Educause Center for Applied Research.

Black, P. and William, D. (1998). "Inside the Black Box: Raising Standards Through Classroom Assessment." *Phi Delta Kappan*, 80(2), October Issue

Chemistry Concepts Inventory. *Journal of Chemical Education*. http://jchemed.chem.wisc.edu/ JCEDlib/QBank/collection/CQandChP/CQs/ConceptsInventory/CCIIntro.html, (accessed July 8, 2009)

Duncan, D. (2005). *Clickers in the Classroom: How to Enhance Science Teaching Using Classroom Response Systems*. San Francisco: Pearson Education, Inc.

Duncan, D. (2006). "Clickers: A New Teaching Aid with Exceptional Promise." *Astronomy Education Review*, 5(1), 70–88

Hake, R. R. (1998). "Interactive engagement versus traditional methods: A six-thousand-student survey of mechanics test data for introductory physics courses," *American Journal of Physics*, 66(1), 64-74

Hoekstra, A (2008). "Vibrant student voices: exploring effects of the use of clickers in large college courses." *Learning, Media, and Technology*, 33(4), 329–341

Mayer, R.E., Stull, A., DeLeeuw, K., Almeroth, K., Bimber, B., Chun, D., Bulger, M., Campbell, J., Knight, A., and Zhang, H. (2009). "Clickers in the classroom: Fostering learning with questioning methods in large lecture classes." *Contemporary Educational Psychology*, 34(1), 51–57

Mazur, E. (1997). *Peer Instruction: A User's Manual*. Prentice Hall

Nurrenbern, S.C. & Pickering, M. (1987). "Concept learning versus problem solving: is there a difference?" *Journal of Chemical Education*, 64(6), 508–510

Wieman, C. et al. (2008). "Clicker Resource Guide: An Instructor's Guide to the Effective Use of Personal Response Systems (Clickers) in Teaching." http://www.colorado.edu/sei/documents/ clickeruse_guide0108.pdf, (accessed July 8, 2009)

Yarroch, W. L. (1985). "Student understanding of chemical equation balancing." *Journal of Research in Science Teaching*, 22(5), 449–459